River Resource Management in the GRAND CANYON

Committee to Review the
Glen Canyon Environmental Studies

Water Science and Technology Board

Commission on Geosciences, Environment, and Resources

National Research Council

National Academy Press
Washington, D.C. 1996

NOTICE: The project that is the subject of this report was approved by the Governing Board of the National Research Council, whose members are drawn from the councils of the National Academy of Sciences, the National Academy of Engineering, and the Institute of Medicine. The members of the committee responsible for the report were chosen for their special competences and with regard for appropriate balance.

This report has been reviewed by a group other than the authors according to procedures approved by a Report Review Committee consisting of members of the National Academy of Sciences, the National Academy of Engineering, and the Institute of Medicine.

Support for this project was provided by the U.S. Department of the Interior, Bureau of Reclamation, under Cooperative Agreement Number 6-FC-40-04240.

International Standard Book Number 0-309-05448-6

Library of Congress Catalog Card Number 95-73313

Additional copies of this report are available from:

National Academy Press
2101 Constitution Avenue, NW
Box 285
Washington, DC 20055
800-624-6242
202-334-3313 (in the Washington Metropolitan Area)
B-719

Printed in the United States of America

Cover art by Larry Stevens, Flagstaff, Arizona

COMMITTEE TO REVIEW THE GLEN CANYON ENVIRONMENTAL STUDIES

WILLIAM M. LEWIS, JR. *(Chair)*, University of Colorado, Boulder
GARRICK A. BAILEY, University of Tulsa, Oklahoma
BONNIE COLBY, University of Arizona, Tucson
DAVID DAWDY, Consulting Hydrologist, San Francisco, California
ROBERT C. EULER, Consulting Anthropologist, Prescott, Arizona
IAN GOODMAN, The Goodman Group, Boston, Massachusetts
WILLIAM GRAF, Arizona State University, Tempe
CLARK HUBBS, University of Texas, Austin
TREVOR C. HUGHES, Utah State University, Logan
RODERICK NASH, University of California, Santa Barbara (through 1994)
A. DAN TARLOCK, IIT Chicago Kent College of Law, Chicago, Illinois

Staff

SHEILA D. DAVID, Study Director
MARY BETH MORRIS, Senior Project Assistant

WATER SCIENCE AND TECHNOLOGY BOARD

The National Academy of Sciences is a private, nonprofit, self-perpetuating society of distinguished scholars engaged in scientific and engineering research, dedicated to the furtherance of science and technology and to their use for the general welfare. Upon the authority of the charter granted to it by the Congress in 1863, the Academy has a mandate that requires it to advise the federal government on scientific and technical matters. Dr. Bruce M. Alberts is president of the National Academy of Sciences.

The National Academy of Engineering was established in 1964, under the charter of the National Academy of Sciences, as a parallel organization of outstanding engineers. It is autonomous in its administration and in the selection of its members, sharing with the National Academy of Sciences the responsibility for advising the federal government. The National Academy of Engineering also sponsors engineering programs aimed at meeting national needs, encourages education and research, and recognizes the superior achievements of engineers. Dr. Harold Liebowitz is president of the National Academy of Engineering.

The Institute of Medicine was established in 1970 by the National Academy of Sciences to secure the services of eminent members of appropriate professions in the examination of policy matters pertaining to the health of the public. The Institute acts under the responsibility given to the National Academy of Sciences by its congressional charter to be an adviser to the federal government and, upon its own initiative, to identify issues of medical care, research, and education. Dr. Kenneth I. Shine is president of the Institute of Medicine.

The National Research Council was organized by the National Academy of Sciences in 1916 to associate the broad community of science and technology with the Academy's purposes of furthering knowledge and advising the federal government. Functioning in accordance with general policies determined by the Academy, the Council has become the principal operating agency of both the National Academy of Sciences and the National Academy of Engineering in providing services to the government, the public, and the scientific and engineering communities. The Council is administered jointly by both Academies and the Institute of Medicine. Dr. Bruce M. Alberts and Dr. Harold Liebowitz are chairman and vice chairman, respectively, of the National Research Council.

"This evening, as I write, the sun is going down, and the shadows are settling in the canyon—the gateway through which we are to enter on our voyage of exploration tomorrow. What shall we find?"

John Wesley Powell, THE EXPLORATION OF THE COLORADO RIVER 15 (1875).

Preface

For about 50 years, culminating in the 1970s, the United States steadily dammed most of its large rivers. By the end of this era, the technical, fiscal, and political evolution of dam building had proceeded to such a refined state that dams could be placed where previously they would have been considered impossible, impractical, or inadvisable. Nevertheless, by 1980 America's Age of Impoundment had closed under a storm of environmental opposition and fiscal criticism.

Remission in the national struggle over approval of new dams has allowed public attention to be redirected toward the operation of existing dams. Most large dams were originally justified by water supply, flood control, and production of hydroelectric power. Accordingly, most dams have operated on an annual or seasonal schedule that reflects demand for storage capacity or water delivery and on a daily or weekly schedule that reflects hourly fluctuations in the value of hydroelectric power. In the meantime, public interests that might previously have been considered distantly secondary or even frivolous have become potentially serious considerations affecting the operation of dams. These include fisheries that were produced incidentally to impoundment, recreational boating or rafting, welfare of aquatic life, protection of culturally significant sites, and even aesthetic preferences. A case in point is the Glen Canyon Dam on the Colorado River.

There was never any doubt that Glen Canyon Dam would change the Colorado River. The dam traps the river's sediment and thus replaces turbid, sediment-laden water with clear water that is hungry for sediment, but many of the environmental effects downstream of the dam were not fully understood at the beginning of the GCES. Water drawn through turbines at great depth is constantly cold, even in the Arizona summer. The seasonal swing of discharge, which originally spanned an amplitude as much as 100-fold, was replaced by a rhythm reflecting hourly changes in the market for electricity. These were the most obvious consequences of the dam. No doubt some experts also foresaw indirect consequences, but until recently these have been poorly documented. They include displacement of native fishes through the chilling of the river, as well as depletion and redistribution of sediment in ways that affect camping and backwaters of importance to aquatic life. The dam has caused physical changes in culturally important sites and has led to the development of a new suite of riparian vegetation. The aesthetic features of the river through the Grand Canyon, although difficult to quantify, also have changed.

Changes in resources other than power and water delivery can potentially be controlled or moderated by adjustments in the operation of Glen Canyon Dam. Operational changes could be used in building and preserving beaches, stabilizing backwaters, improving the propagation of native fishes, protecting cultural sites, and optimizing the aesthetic experience of visitors to the canyon. Even so, these possibilities present complications, including sacrifice of power revenue, potential conflict among optimal operating regimes for different resources, and unintended consequences less desirable than the status quo.

The scarcity of information on environmental resources has slowed consideration of alternative operating schemes for Glen Canyon Dam. The Bureau of Reclamation (BOR) took a significant step in acknowledging the need for information when it authorized the Glen Canyon Environmental Studies (GCES) in 1982. Although at first narrowly centered around Glen Canyon Dam (hence the specific reference to it), the GCES expanded as it became clear that the operation of Glen Canyon Dam directly affects numerous environmental resources along the more than 250 miles of the Colorado River between Glen Canyon Dam and Lake Mead. In this way, GCES became the vehicle by which the river corridor was first recognized as an integrated environmental system that responds to the operation of Glen Canyon Dam. Quite apart from the information that it has produced, GCES has been important in redefining the scope of responsibility for management of Glen Canyon Dam.

The purpose of this report and the committee's task has been to review research that has been done in connection with the Glen Canyon Environmental Studies and to comment on the application of science in the management program for the Colorado River. Perceiving the need for independent review and oversight of the GCES, the BOR requested the formation of a National Academy of Sciences/National Research Council (NRC) review committee through the NRC's Water Science and Technology Board. The committee began its work in 1986 and has continued through several phases of reauthorization to the present. Since 1986, it has made numerous recommendations, some of which have affected the design of GCES. The committee has considered not only the technical aspects of environmental studies, which will affect the future operation of Glen Canyon Dam, but also the broader significance of GCES as an example of large governmental ecosystem studies. This report reviews GCES reports received by the committee through July 1995.

Strong political and institutional forces meet in the Grand Canyon. The present report deals with technical issues but also attempts to describe as completely as possible all of the factors that explain the successes and failures of the GCES program. The report finds many shortcomings in GCES. In identifying and analyzing them, the committee has been increasingly aware that problems associated with GCES have, in large part, been a reflection of the federal government's lack of experience in conducting studies that deal comprehensively with many kinds of resources in an ecosystem context. In addition, the committee has seen that most of the deficiencies in GCES derive from the organizational culture of federal agencies, which are not well acclimated to easy collaboration with each other or with external scientific and technical communities. Many of the problems brought out by the committee's analysis of GCES are not attributable to the individual project participants. In fact, one irony of GCES is that it has benefited from the energies of numerous remarkably dedicated and knowledgeable individuals but has still shown major flaws that were essentially beyond the control of the individual participants.

Despite its flaws, GCES has been the catalyst for major changes in the operation of Glen Canyon Dam. In fact, if seen as an example of interaction between science and environmental resource management, GCES is an extraordinary success. The BOR deserves much credit for adapting its management practices to new knowledge of the environmental system as produced through GCES.

The committee's work on GCES has extended over a far longer interval than most NRC committee projects. The committee benefited enormously from the efforts of Sheila David, the NRC study director. Throughout its nine years of operation, she provided the thread of continuity that maintained the focus and purpose of the committee. She ensured sound management of the committee's resources, both financial and intellectual, and provided a rational voice in all committee debates. Her intelligence and skill are an integral part of every product of the committee. In addition, the committee thanks Mary Beth Morris, WSTB project assistant, for all her help during committee meetings and with the production of this report. The NRC committee and the BOR have benefited from the guidance and assistance of GCES project manager David Wegner, whose investments of time, energy, and intellectual interest in the research being conducted in the GCES program have been invaluable. The committee also received much assistance from numerous other individuals of the NRC staff and of the cooperating agencies and the Native American tribes that kept the committee informed and encouraged its work.

William M. Lewis, Jr.
Chair
Committee to Review the Glen Canyon Environmental Studies

Contents

Executive Summary

The Colorado River corridor through the Grand Canyon is among the most highly valued natural resources of the United States. Although the Grand Canyon corridor is a national park and Glen Canyon above it is a national recreation area, the flow of the Colorado River through both of these has been regulated for water management and power production since 1963, when the Glen Canyon Dam was installed near Page, Arizona. Glen Canyon Dam offers many benefits, including the measured distribution of approximately 15 million acre-feet of valuable western water, the generation of hydropower valued at $50 million to $100 million per year, the maintenance of a prized coldwater trout fishery below the dam, the control and storage of sediment, and recreation in Lake Powell. Among the environmental costs of the dam, however, are the suppression of native fishes, including endangered fishes native to the Colorado River, erosion of beaches valued as campsites by rafters, support of aquatic and terrestrial organisms, and large daily changes in discharge volume and water level that are potentially harmful to aquatic and riparian communities and are considered aesthetically undesirable by most visitors to the river.

The daily and seasonal operating regimes of Glen Canyon Dam were challenged in the early 1980s by constituencies calling for the moderation of operating regimes in recognition of values other than hydropower production. Responding to these challenges, the Bureau of Reclamation (BOR) initiated in 1982 the Glen Canyon Environmental Studies (GCES), which were intended to document the effects of dam operations on resources other than hydroelectric power. In committing itself to analyze the effects of dam operations on all resources downstream of Glen Canyon Dam, the BOR undertook some of the most ambitious integrated environmental studies ever conducted under federal sponsorship. The GCES, which extended over an interval of 13 years at a cost of over $50 million, have produced information leading to major changes in the operation of Glen Canyon Dam. The GCES program is important not only in rationalizing changes in the operation of Glen Canyon Dam, but also as a forerunner and prominent example of the challenges that federal agencies face in conducting complex environmental studies that ultimately allow the comparison of costs and benefits across diverse categories of resources.

WORK OF THE NRC COMMITTEE

In 1986, after Phase I of the GCES was underway, the BOR requested that the National Academy of Sciences appoint, through the National Research Council's Water Science and Technology Board, a committee to oversee and review the GCES. The committee was formed in 1986 and continued its work through the end of GCES in 1995. Between 1986 and 1995, the NRC committee released 12 reports related to GCES (see page 8). These included a review of the first phase of GCES (NRC, 1987), a review of the environmental impact statement (EIS) on dam operations (NRC, 1994a), and a review of the draft federal long-term monitoring plan (NRC, 1994b). In addition, the NRC committee convened a symposium to assist the GCES senior scientist in gathering background information on the Colorado River (NRC, 1991a), sponsored a workshop in 1993 that was to form the basis for a long-term monitoring plan, and sponsored two meetings with the six Native American tribes that were participants in the GCES. The committee's task has been to review research that has been done in connection with the Glen Canyon Environmental Studies and to comment on the application of science in the management program for the Colorado River.

OBJECTIVES AND DESIGN OF THE GCES

The objective of GCES was to identify and predict the effects of variations in operating strategies on the riverine environment below Glen Canyon Dam within the physical and legal constraints under which the dam must operate. Critical elements for the development of GCES and other such projects include a list of resources directly or indirectly affected by management, a list of management options, and an ecosystem framework showing the causal connections among system components, potential management strategies that include humans as integral parts of the environment. The GCES ultimately developed each of these elements, but only during the course of study rather than prior to it. Future studies could benefit from earlier definition of scope.

The GCES showed how agency perspectives and their legally defined missions can constrain a list of management options. Because the leadership of the GCES program lacked independence and authority within BOR, the valid objectives of other agencies involved in the research did not always receive adequate attention (NRC, 1988b). The BOR, with support from the U.S. Department of Energy's Western Area Power Administration (WAPA), initially argued that flexibility in the operation of Glen Canyon Dam did not extend to any operating regime that would reduce hydropower revenues. The BOR subsequently changed its policy in this regard, thus opening the scope of GCES to a more effective breadth. Future studies should allow definition of management options that occupy the full range of possibilities rather than the preferred operating range of the sponsoring agency.

The ecosystem concept is essential in unifying studies of environmental systems. While GCES Phase II recognized the importance of the ecosystem concept, the components of GCES already had been identified and were never fully realigned with the ecosystem approach.

RESULTS OF THE GCES

Although GCES produced numerous useful results, it had not completed any final synthesis or integrated report as of September 1995. It is not clear whether final synthesis will occur, nor was the NRC committee able to review such a synthesis even in draft form. Firm commitments to final synthesis and specific recommendations to management should always be required for future studies of this type. Unless there are compelling reasons not to do so, the synthesis should be completed prior to the time that management decisions are finalized. Final synthesis should reflect uncertainties that remain at the end of the project and show how resource managers can accommodate uncertainty through adaptive management.

This section summarizes the general conclusions and recommendations made by the committee. Other more detailed recommendations can be found at the end of Chapters 4, 5, 8, 9 and 10. Chapter 11 summarizes the lessons learned from the committee's involvement in review of the GCES and provides suggestions to the BOR and other government agencies for future studies of complex environmental systems.

Operations of Glen Canyon Dam

The BOR, through GCES, was responsible for a number of significant achievements between 1982 and 1995. Information from GCES on the linkage between operations and the transport and distribution of sediment below the Glen Canyon Dam supported specific recommendations for change in operation of the dam. Equally as significant, the BOR facilitated major operational changes in recognition of this new information from GCES and committed itself to the concept of adaptive management, which will involve frequent consideration of adjustments in operations as a means of optimizing the aggregate value of all resources below the dam.

For the future, the scope of adaptive management can be extended through installation of a multiple-level outlet withdrawal and possibly by other means as well. Options that now have little support but that may offer some significant advantages need to be explored objectively. These include slurry pipelines for augmentation of sediment supply and a reregulation dam that would allow more complete control of flow for environmental purposes while also allowing maintenance of maximum power revenues.

Sediment and Hydrology

The GCES provided valuable information on sediment dynamics in the Colorado River between Glen Canyon Dam and Lake Mead. The GCES verified that the supply of sand reaching the Colorado River through tributaries below Lee's Ferry is sufficient to maintain beaches between Lee's Ferry and Lake Mead. GCES also showed that beaches are best protected by operating regimes that avoid great extremes in daily discharge or in the rate of change in daily discharge. Even with protective measures in the form of less extreme daily fluctuations, as adopted by BOR in response to GCES, gradual loss of mass from beaches and gradual sedimentation in backwaters and pools will occur unless occasional high discharges (controlled floods or beach-building flows) are part of the operating regime. These high flows lift sand from pools in the river channel onto the beaches and scour backwaters.

GCES also showed that such high flows are critical in moving coarse debris that enters the Colorado River with storm flows from tributaries. Thus, GCES showed the justification for a new operating regime, and the BOR recommended adoption of this new operating regime through the EIS on the operation of Glen Canyon Dam. The GCES findings on sediment and the BOR's application of these findings are exemplary of an effective interface between environmental analysis and resource management. The new operating regime will result in moderate losses of power revenue but will offer substantial benefits to recreational interests (rafting, fishing); aquatic life, including endangered species; and nonuse value.

The committee recommends several kinds of continuing research and monitoring that will enhance the potential for beneficent management of sediment transport:

- Study of the rate of sand interchange between the main channel and the eddy systems that create beaches.
- Development of a mechanism for determining the initiation of beach-building flows.
- Development of quantification of the magnitude and duration of beach-building flows.
- Study of the rate at which sand is deposited on beaches during beach-building flows.
- Creation of a procedure for determining sand budgets in different parts of the canyon downstream from Glen Canyon Dam.

Biological Resources

The GCES analysis of biotic components also produced valuable new information, but was not well integrated. The GCES provided some of the first comprehensive inventories of aquatic life along the Colorado River corridor and resulted in an excellent study of the humpback chub and studies of other endangered species and of trout. While this information is essential in support of ecosystem analysis, GCES failed to progress to a comprehensive view of connections between biotic components, physical or chemical habitat features, and operations. Synthesis was notably absent, and predictive capability was weak. In the future, careful planning that is focused on specific objectives known in advance to be useful to management will be critical, as will commitment to completion of the advanced phases of ecosystem analysis.

Recreation

GCES-supported studies of recreation quantify and explain the changes in economic value associated with rafting under various strategies for operation and management of the dam. GCES also provides predictions related to the trout fishery, but these predictions have an inadequate basis because of the weak understanding of the composite effects of various factors on the trout population.

Hydropower Economics

Hydropower economics were studied extensively in the late phases of GCES. These studies were strengthened by the inclusion of external experts in the improvement and constructive critique of modelling for the purpose of projecting the costs of various operational alternatives. Only through studies of this type can the environmental benefits of various operating alternatives for the dam be weighed against hydropower production. For the future, the BOR should sustain these gains by developing analytical modelling capabilities for adaptive management. BOR should maintain a national economic perspective in its projections, and should consider costs of short-term flow alternatives in the context of potential long-term benefits.

Cultural Resources

Cultural resources, which include both sites of cultural significance (historic and prehistoric) and sites of tribal significance not marked by specific artifacts, were originally not accorded appropriate recognition in the GCES study plan. In the late phases of GCES, some of these components, as well as some studies of endangered species, were examined in a manner that did not always address the basic objectives of GCES. Future projects of a similar type should take into account cultural resources from the early planning stages and should constrain studies around the objectives of the project.

Institutional Influences on the GCES

GCES, although successful in a number of respects, was handicapped by organizational and administrative flaws that are deeply imbedded in the operating traditions of the federal government. In organizing GCES, the BOR properly convened a group of cooperators consisting of federal agencies with research interests in the Colorado River, a state agency, and, belatedly, Native American tribes having cultural and resource interests in the Colorado River between Glen Canyon Dam and Lake Mead. The GCES project manager was not, however, vested with sufficient authority to override the influence of individual cooperators for the overall benefit of the project. One result was expansion of the scope of work as a reflection of specific agency interests but not always according to the priorities of GCES. Agency personnel also developed strong feelings of entitlement to GCES support, which they reinforced through control over facilities and permitting. This greatly reduced the possibility for diversification of expertise through open solicitation of proposals outside the federal government. The development of agency entitlements also weakened the ability of project management to enforce contractual obligations made by agencies to GCES and increased project costs by diffusing the focus of study. These institutional problems are very likely generic to cooperative ventures involving federal agencies and need to be remedied in future projects through greater independence and authority of project management over project resources.

Nonuse Values

Nonuse values (values not associated with direct use of the resource) are exceptionally high for the Grand Canyon region because of its aesthetic and cultural appeal regionally, nationally, and internationally. Even so, the BOR resisted for most of the history of GCES any inclusion of nonuse values as a consideration related to operational alternatives. The BOR did, however, ultimately support through GCES a state-of-the-art study of nonuse values, which was appended to the EIS in 1995. Studies of nonuse values show that distinctions among operating alternatives for Glen Canyon Dam are perceived as significant by the public and that the U.S. public's willingness to pay for preferred operating alternatives for the dam is of the order of several billion dollars. This is many times the value of changes in any of the tangible resources affected by the dam's operations, including power production. While the implications of this information for operations are as yet uncertain, their relevance to a broad perspective on operations is unquestionable. Thus, GCES has illustrated the need for inclusion of nonuse value studies in similar projects.

ACHIEVEMENTS OF GCES

The BOR adjusted the scope of GCES inquiry in a manner consistent with the ecosystem concept, thus encompassing, ultimately, all resources and a full range of management options as the basis for analysis of the effects of dam management. While this adjustment came too late to be fully effective for GCES, the conceptual advance itself is highly significant and should be carried forward to other projects. The BOR also developed, over the course of GCES, greatly increased acceptance of external criticism and expertise through its work with the NRC committee as well as its appointment of an external senior scientist for GCES and through formation of an advisory board, increased emphasis on external contracting, and acceptance of external participation in hydropower studies. The BOR has, through GCES, modernized and reformed its strategy for management of Glen Canyon Dam. Similar changes, augmented by lessons from GCES, could inaugurate a new era in the management of western waters by BOR and other federal agencies.

LIST OF NRC REPORTS

National Research Council. 1987. River and Dam Management: A Review of the Bureau of Reclamation's Glen Canyon Environmental Studies. Washington, D.C.: National Academy Press.

National Research Council. 1988a. Supplementary Report to 1987 River and Dam Management. Water Science and Technology Board, National Research Council, Washington, D.C.

National Research Council. 1988b. Letter report to the Honorable Donald Paul Hodel, Secretary, U.S. Department of the Interior, August 1, from G. Richard Marzolf, Chair, Committee to Review the Glen Canyon Environmental Studies. Water Science and Technology Board, National Research Council, Washington, D.C.

National Research Council. 1991a. Colorado River Ecology and Dam Management: Proceedings of a Symposium, May 24-25, 1990, Santa Fe, N. Mex. Washington, D.C.: National Academy Press.

National Research Council. 1991b. Review of the Draft Integrated Research Plan for the Glen Canyon Environmental Studies, Phase II. Washington, D.C.: National Academy Press.

National Research Council. 1991c. Letter report to Commissioner Dennis Underwood, August 9, from William M. Lewis, Jr., Chair, Committee to Review the Glen Canyon Environmental Studies. Water Science and Technology Board, National Research Council, Washington, D.C.

National Research Council. 1991d. Letter report to the Honorable Manuel Lujan, Office of the Secretary, U.S. Department of the Interior, August 9. Water Science and Technology Board, National Research Council, Washington, D.C.

National Research Council. 1992a. Letter report to Michael Roluti, Bureau of Reclamation, October 21, Committee to Review the Glen Canyon Environmental Studies comments on May 1992 draft report "Power System Impacts of Potential Changes in Glen Canyon Power Plant Operations." Water Science and Technology Board, National Research Council, Washington, D.C.

National Research Council. 1992b. Letter report to David L. Wegner, April 22, assessing proposed GCES studies related to economics, hydropower production, and dam operations. Water Science and Technology Board, National Research Council, Washington, D.C.

National Research Council. 1993. Letter report to Tim Randle, Bureau of Reclamation, February 26, Committee to Review the Glen Canyon Environmental Studies comments on January 1993 preliminary draft "Operation of Glen Canyon Dam, Colorado River Storage Project, Arizona." Water Science and Technology Board, National Research Council, Washington, D.C.

National Research Council. 1994a. Review of the Draft Environmental Impact Statement on Operation of Glen Canyon Dam. Washington, D.C.: National Academy Press.

National Research Council. 1994b. Review of the Draft Federal Long-Term Monitoring Plan for the Colorado River Below Glen Canyon Dam. Washington, D.C.: National Academy Press.

1

Introduction and Background

In writing about the Colorado River, John Wesley Powell (1895) noted the irony of snowmelt maintaining a river through hundreds of miles of the hottest and driest terrain in North America. Descending from almost 3 miles above sea level near Long's Peak in Colorado and in the Wind River range of Wyoming, the Colorado River's sources quickly reach flatter topography in Utah and Colorado while still more than a mile above sea level. The river continues its plunge toward sea level through canyons that ultimately trench the plateau to a depth of more than a mile. Because the canyons allow the river to seek its final level quickly, the Colorado moves with notorious energy.

A modern irony equal to Powell's irony of ice water in the desert is that the Colorado River, known for its extraordinary rush to the sea, now typically fails to reach the sea at all. In Powell's day, water moved freely from the snow fields of the Rockies to the Gulf of California in 1 to 3 months, depending on the season. At present, only about 10 percent of the annual yield crosses from the United States to Mexico, where even this remnant is consumed before it can reach the Gulf of California. Although Powell nodded approvingly at evidence of prehistoric irrigation practices by Indian tribes along the Colorado (Powell, 1895), he would probably have been surprised to see the entire flow of the Colorado thus fully diverted for human use.

The Colorado's flow is held by two very large reservoirs and many smaller ones, all of which comprise a total volume of 58 million acre-feet (Andrews, 1991), or approximately four times the mean annual flow of the river. Water passes from the reservoirs for consumptive use, most of which is accounted for by irrigation. As a significant by-product, the reservoirs of the Colorado River in aggregate produce about 13 billion kwh of hydroelectric energy

(Plummer, 1983).

Of the many reservoirs that impound the Colorado, Lakes Powell and Mead are the most impressive; they are the two largest reservoirs in the western United States. The dams for these two reservoirs, which produce most of the electricity and hold most of the water of the Colorado River, together inundate over 300 miles of the river. While the two dams are of similar size and provide many of the same benefits, Glen Canyon Dam, which impounds Lake Powell, is of particular importance because it holds back the Colorado River in front of the Grand Canyon (Figure 1.1).

Congress authorized the construction of Glen Canyon Dam in 1956 as part of the Colorado River Storage Project Act. The dam was proposed by the Bureau of Reclamation (BOR), which assumed responsibility for its operation when the dam closed in 1963. The Glen Canyon Dam was a realization of the Reclamation Act of 1902, according to which arid lands of the American West would be extensively irrigated with waters from western rivers. Without a dam on the upper mainstem of the Colorado River, use of its water would have been inefficient and thus contrary to the goals of reclamation.

Two factors added particular motivation to the construction of a dam on the upper mainstem of the Colorado River. First was the Law of the River, which includes not only the Colorado River Compact of 1922 but also all previous and subsequent statutes, judicial decisions, and treaties that affect the disposition of the river's waters (Ingram et al., 1990). The Law of the River evolved under the assumption that the Colorado River could deliver at Lee's Ferry just over 16 million acre-feet per year, of which 15 million was to be divided equally between the states of the upper basin (Colorado, Utah, Wyoming) and the lower basin (California, Arizona, New Mexico, Nevada) and the remaining 1.5 million was to go to Mexico. By 1950, it was clear that this assumption was based on an overestimate, which, when combined with inevitable droughts, would either prevent the upper-basin states from using their full allocation or would cause them to default on their delivery obligations to the lower basin (Dawdy, 1991). A large dam on the upper mainstem would cover the immediate damage from miscalculation of the yield until such time as an extended drought might coincide with the upper basin's full use of its allocation. In addition, a dam on the upper mainstem would produce hydroelectric power revenues that could be used to finance delivery or storage of water throughout the basin (Ingram et al., 1991). Without subsidy from hydropower, the BOR's constellation of water projects in the Colorado River basin would have been impractical, especially in the sparsely populated

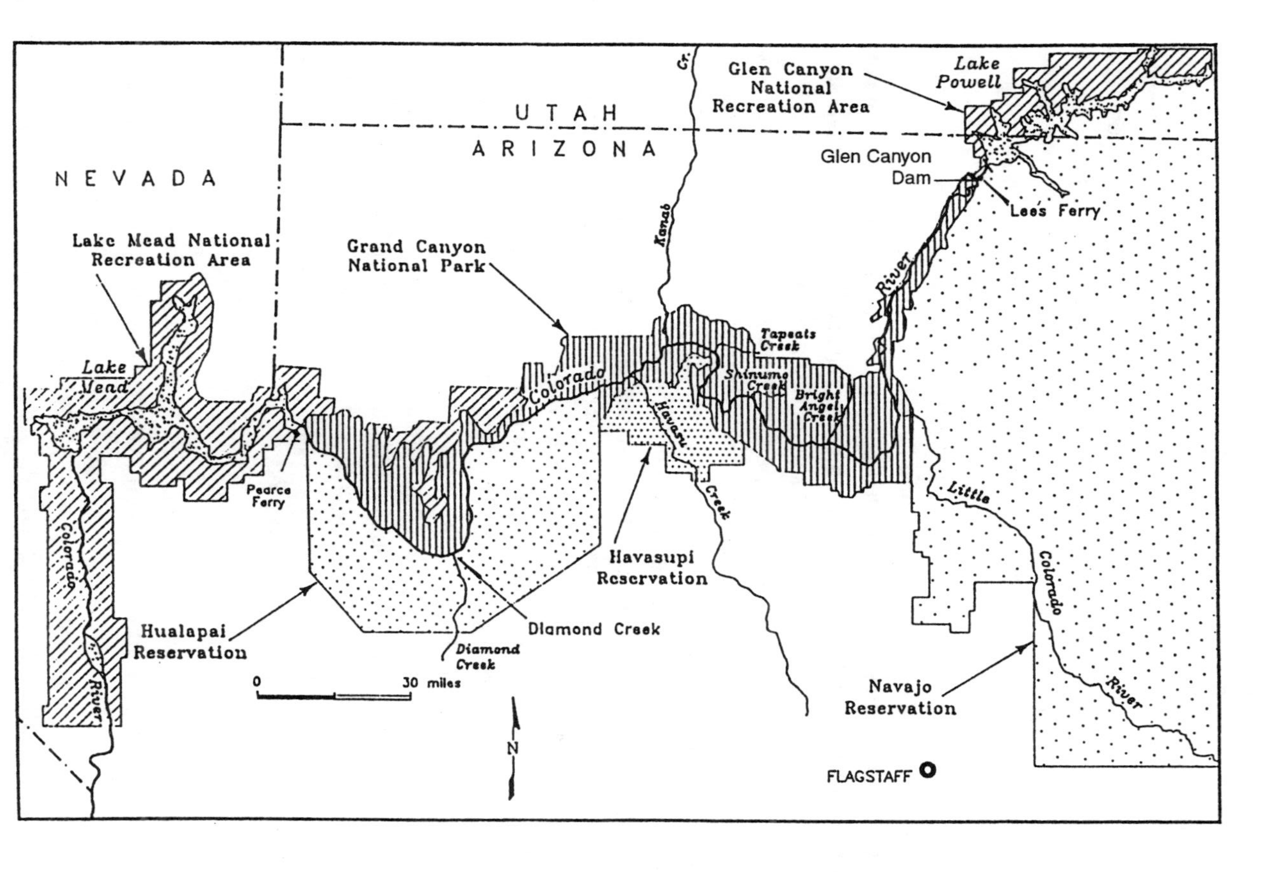

FIGURE 1.1 General area map of the Grand Canyon. SOURCE: Redrawn from Valdez and Ryel (1995).

upper basin, because most projects required strong external subsidies that could be sustained only by a revenue stream. Thus, the Glen Canyon Dam was not merely a reclamation project but also a guarantor of integrity for the Law of the River and patron of regional reclamation projects that otherwise would have perished for lack of cash.

The Colorado River Storage Project Act listed beneficial use of water, reclamation of arid and semiarid lands, and control of floods as primary purposes of the Glen Canyon Dam. The act also mentioned hydroelectric power but only as an "incident" of other specifically mentioned purposes. In qualifying hydropower production in this way, the act ensured that power production would never challenge the Law of the River. Even so, Congress showed its concern for production of revenue by specifying that Glen Canyon Dam be operated in such a way as to produce maximum power at firm rates consistent with the overriding requirements for delivery of water.

In practice, the secondary status of hydropower proved to be not very restrictive because the central requirement for efficient hydropower marketing is flexibility in hourly or daily scheduling of discharges, whereas such short time scales are typically irrelevant to the delivery of water for consumptive use (Hughes, 1991). Power production and water delivery might come into conflict but only under the most extreme hydrological conditions. Thus, Glen Canyon Dam was able to deliver water, power, and money in quantity by using an operating regime that featured annual divisions of runoff between the upper— and lower— basin states and daily or hourly phasing of discharge in response to fluctuations in demand for power.

Glen Canyon Dam aroused only minor opposition in the 1950s when it was authorized. Most of the opposition came from the Bureau of Reclamation which initially rejected the site because of the problems caused by the soft sandstone. The site was a compromise selection in 1956 after conservationists succeeded in stopping a dam on the Green River which would encroach on Dinosaur National Monument and traded Glen Canyon for no dam on the Green (Fradkin, 1981.) In retrospect, it is clear that Glen Canyon Dam slipped through a partly closed door; two other mainstem projects (Marble Gorge and Bridge Canyon) running slightly behind Glen Canyon were killed by public opposition.

Struggles over the authorization of new dams were so intense from 1965 to 1980 that they diverted attention from the operation of existing dams. Management principles for Glen Canyon Dam were essentially static for almost 20 years following completion of the dam, even though the political landscape metamorphosed drastically over this interval. A changed perception of the public welfare was evident as far back as 1968, when the Colorado River Basin Project Act broadened the definition of purposes for operation of the dam to include not only storage and delivery of water but also maintenance of water quality, outdoor recreation, and fish and wildlife. The 1968 act again listed power production as an incident of other purposes. The broadened purposes for operation of the dam were reconfirmed in 1992 with the Grand Canyon Protection Act, which cited the need to "mitigate adverse impacts to and improve the values for which Grand Canyon National Park and Glen Canyon National Recreation Area were established, including, but not limited to natural and cultural resources and visitor use."

Several forces were behind Congress' decision to intervene in the operation of Glen Canyon Dam. The most important change in the legal frame of reference was the Endangered Species Act (ESA) of 1973. Although the ESA did not immediately change the operation of Glen Canyon Dam, it set the stage for operations that would be answerable to the welfare of the Colorado River's endemic fishes and possibly to endangered species of the riparian zone as well. The effect of the ESA was magnified by the National Environmental Policy Act (NEPA) of 1970, which assured an extended public examination of the welfare of endangered species and of other resources that had not been considered in the development of the original operating rules for the dam. Given that the NEPA required significant federal action as a trigger, however, it did not begin to affect the operation of Glen Canyon Dam until the 1980s.

New constituencies also developed around Glen Canyon Dam. The trout fishery of the dam's tailwater maintained a steady concern for the welfare and growth of trout, while the rafting industry, which is based on some 30,000 annual participants in short trips and 20,000 participants in long trips through the canyon (BOR, 1993), focused attention on the number, size, and stability of beaches for camping. Following judicial validation of their rights to the river's water, Native American tribes gradually also asserted their claim to evaluate the dam's operations. Finally, environmental groups, representing not only their members but also a more diffuse change in societal attitudes about environmental resources, showed their willingness to discard the initial assumptions upon which the operating rules were based.

The traditional operating rules were of direct benefit to the consumers of power through the regional marketing network. Power from Glen Canyon Dam, although produced by the BOR through management of the dam, is marketed by the Western Area Power Administration (WAPA), an agency of the U.S. Department of Energy. Given that the daily and weekly operating schedules for the dam had been optimized around the marketing of power, any change in the operating rules could be expected to reduce the value of power from Glen Canyon Dam. WAPA spoke against changes in the operating rules on these grounds. The BOR was predisposed to identify strongly with WAPA's priorities. Not only had the bureau been directed in the original authorizing legislation for Glen Canyon Dam to produce maximum revenue from power, it had also assumed that revenue from the dam would sustain its other projects in the Colorado River Basin.

Between 1970 and 1980, major changes in the operating rules for Glen Canyon Dam would have been difficult, even if the BOR had wanted to make them. Water delivery, flood control, and optimization of hydropower revenues presented a complex but deterministically soluble set of demands for release of water from the dam. In contrast, the requirements of endangered species, trout, recreation, and Native American tribes could not be specified even qualitatively because they had not been sufficiently studied. Prior to 1980, the BOR could find no basis for changing the operating rules of Glen Canyon Dam, even though an array of new resources had been added to the list of purposes for operating the dam.

Before passage of the NEPA, federal management agencies had legitimately claimed that they were not authorized to study the environmental effects of management practices. NEPA reversed this line of reasoning overnight; studies were not only allowed, they were mandatory.

In the early 1980s the BOR foresaw the need to rewind the Glen Canyon generators and also proposed to increase generator capacity and adjust operations to increase output of peaking power. This led it to consider the preparation of either an environmental assessment or a full environmental impact statement (EIS). The bureau hoped to avoid an EIS, which would have been the first to examine the full scope of operations for a reservoir constructed prior to passage of the NEPA. To support its decision not to conduct an EIS the BOR needed data. To provide the data, it created the Glen Canyon Environmental Studies (GCES).

The bureau at first constrained the GCES very narrowly, partly on the incorrect theory that the Law of the River precluded significant operational changes. Environmental data were to be collected primarily near the dam, and analysis of operating regimes that might result in significant loss of efficiency in hydropower marketing was prohibited. Analysis of aesthetic and cultural values was also ruled out of bounds. While needing to provide support for its contention that a small increase in generator capacity would have at most small environmental effects, the BOR clearly did not wish to invite questioning of its fundamental operating rules. Rather than coming to a quick conclusion, however, the GCES merely proved with increased certainty the need for environmental studies of broader scope.

In administering and designing GCES, the BOR maintained its close working relationship with WAPA, which advised it on potential impairment of hydropower marketing that might result from changes in operation of the dam. In fact, the revenues gathered by WAPA from hydropower paid for the GCES program. In addition, the BOR created an interagency group consisting of agencies with responsibility for protection or management of environmental resources below Glen Canyon Dam. The agencies initially included the National Park Service, the U.S. Fish and Wildlife Service, and the Arizona Game and Fish Department but were later broadened to include Indian tribes with territorial or cultural interests in the Grand Canyon. Much of the push to extend and broaden GCES came from the resource management agencies that, without the BOR, had no means of financing extensive studies of the resources below Glen Canyon Dam.

The BOR was criticized externally for lack of credibility in conducting environmental studies. Although agencies with environmental responsibilities were well represented in the design and execution of the GCES, the bureau, in cooperation with WAPA, provided both the money and management for GCES. Given the BOR's historical commitment to water delivery and power production, its control of GCES presented at least the potential for conflict of interest.

ROLE OF THE NRC

Desiring to maintain credibility for its newly expanded studies, the BOR in 1986 requested that the National Research Council (NRC) form a committee to provide scientific reviews of the GCES. The NRC's Water Science and Technology Board sponsored the formation of a committee in 1986 (Table 1.1).

The first task of the Committee to Review the Glen Canyon Environmental Studies was to review the results of Phase I of GCES, which was nearing completion as the committee was being formed. The committee found considerable value in GCES but also criticized the program for insufficient geographic and conceptual scope, weak integration of components, lack of a clear master plan, and excessively internalized staffing of the research effort (NRC, 1987). The committee called for external scientific oversight, inclusion

TABLE 1.1 Sequence of Events for GCES

Event	Year
1. Initiation of GCES Phase I	1983
2. NRC Committee on GCES appointed	1986
3. Release of first NRC report	1987
4. Initiation of GCES Phase II	1987
5. NRC sponsored review of USGS sediment study plan	1988
6. Initiation of environmental impact study	1989
7. Creation of federal cooperators group	1990
8. NRC sponsored Grand Canyon symposium	1990
9. Start of experimental flows for Glen Canyon Dam	1991
10. Start of interim flows for Glen Canyon Dam	1991
11. Addition of Indian tribes to cooperators group	1991
12. Formation of GCES external advisory board	1993
13. NRC review of the long-term monitoring plan	1994
14. Release of U.S. Fish and Wildlife Service Final Biological Opinion on the humpback chub	1994
15. NRC review of draft EIS	1995
16. Release of final environmental impact statement	1995
17. Record of decision for EIS	1996[a]

[a]Anticipated.

of a senior scientist on the study team, more open contracting procedures, better planning and integration, and extension of scope.

The committee had two objectives in making its final review of GCES. The first was to determine the degree to which GCES has been successful in providing a firm scientific basis for assessing the consequences of various management options for Glen Canyon Dam. The second objective was to extract from GCES some of the possibilities and limitations for cooperative government studies of ecological systems that must be managed for the use and protection of multiple resources.

From its review of GCES Phase I, the NRC committee concluded that the GCES program was in part handicapped by the absence of a comprehensive review of environmental studies in the Grand Canyon. As a means of assisting GCES in drawing together all useful information and also of soliciting analysis and comment from scientists with a working knowledge of the Colorado River in the Grand Canyon, the committee sponsored a symposium in May 1990 in Santa Fe, New Mexico. The proceedings of this symposium were published along with a revised set of recommendations for GCES (NRC, 1991). In this second set of recommendations, the committee called for a plan for long-term research and monitoring that would succeed GCES. The committee also recommended that operation of the dam be viewed more flexibly and that a more explicit ecosystem approach be used in analyzing environmental resources. In addition, it repeated its earlier recommendation that scientific expertise external to the cooperating agencies be used much more extensively.

In 1989 Secretary of the Interior Manuel Lujan notified the BOR that future operation of Glen Canyon Dam should be designed on the basis of an EIS to be completed no later than 1994. This requirement was later embodied in the Grand Canyon Protection Act of 1992, which states:

> Long-term monitoring of Glen Canyon Dam shall include any necessary research and studies to determine the effect of the Secretary's actions under section 1204 (c) on the natural, recreational, and cultural resources of Grand Canyon National Park and Glen Canyon National Recreation Area.

The secretary also directed that a second EIS be conducted of the power marketing for Glen Canyon Dam. Preparation of the operations EIS was to be directed by the BOR, and preparation of the marketing EIS was to be directed by WAPA.

The secretary's decision had a profound effect on GCES and on the scope of the NRC committee's work. GCES served as the main source of information for the EIS on operations. Although GCES was not charged with actual preparation of the EIS, time and effort of the GCES team were consumed in providing information for the EIS team. Thus, the EIS slowed the progress of GCES. The NRC committee included review of the draft EIS as part of its work, given that the EIS was an important application of information from GCES (NRC, 1994).

The two EISs were significant in several ways. They set a deadline for firm decisions on the mode of operation for Glen Canyon Dam. In addition, they brought about the application of GCES information to management of the dam and thus set the stage for adaptive management of dam operations. Finally, they marked the end of an era of intensive study and a transition to long-term monitoring.

The GCES was complicated by a number of management and policy issues that required resolution prior to its termination. The first of these was design of an interim flow regime that would bridge the gap between historical operating practices and potentially new operating practices derived from the EIS Record of Decision. The interim flows are an important landmark in the history of Glen Canyon Dam because they signify the decision of the BOR to modify its operation of the dam in a significant way to protect environmental resources, at the cost of reduced efficiency in power marketing. In evaluating interim flows, GCES scientists had the important responsibility of recommending operational strategies that would be the most likely to protect or optimize environmental resources over the short term.

Another important addition to the central task of GCES was the design of a long-term monitoring program. The bureau concluded that the GCES team would be the best source of recommendations on long-term monitoring and elected to include a long-term monitoring plan in the operations EIS. The GCES team turned to the NRC committee for assistance in organizing a workshop on long-term monitoring. The workshop included 50 scientists chosen for their knowledge of the environmental system of the Grand Canyon or for their experience with long-term environmental monitoring in general. The proceedings of the workshop were given to the GCES senior scientist as a basis for design of the monitoring plan. The committee subsequently reviewed the draft monitoring plan and offered a number of suggestions and criticisms, the most important of which were lack of specificity in the plan and failure of the plan to include guidelines for administration or funding of the monitoring program.

The BOR also asked the GCES team to make recommendations for studies to be completed in the event of a controlled flood flow that was tentatively scheduled for early 1994. This flood flow for 1994 was cancelled because of possible lawsuits and because there was not enough water available. An experimental flood flow was then scheduled for spring of 1995 which was cancelled again because of possible lawsuits by the upper basin states and the need for compliance with the National Environmental Protection Act (NEPA). Another experimental flood flow is tentatively scheduled for spring, 1996. Controlled floods had appeared as a component of the preliminary draft EIS and as recommendations from experts on sediment transport, who pointed out that the rebuilding of steadily eroding beaches in the canyon would require periodic high flows of relatively short duration (several days), even though such flows had been avoided in the past because they involve the loss of hydropower revenues. An abundance of water in 1993, combined with the development of a long-term operational plan, set the stage for a controlled flood. GCES was charged with designing a data collection system that would show both the effectiveness of a flood flow in rebuilding beaches and any unexpected negative effects that the flow might have on other resources in the canyon.

The final products of GCES include some 65 individual reports that summarize the studies of individual environmental components such as trout populations, cultural resources, and sediment transport (Figure 1.2). The products of GCES were also initially scheduled to include synthesis of individual studies and use of the synthesis to project the effects of various operating strategies on environmental resources below Glen Canyon Dam. The synthesis was not completed as of September 1995, but the operations EIS includes preliminary synthesis based on information from GCES as of about 1993. The chapters that follow give the NRC committee's review of the GCES.

While the results of the GCES are of pressing relevance to the protection of resources in the Grand Canyon, they also serve as an excellent case study of the federal government's efforts to use ecosystem science as a guide to environmental management.

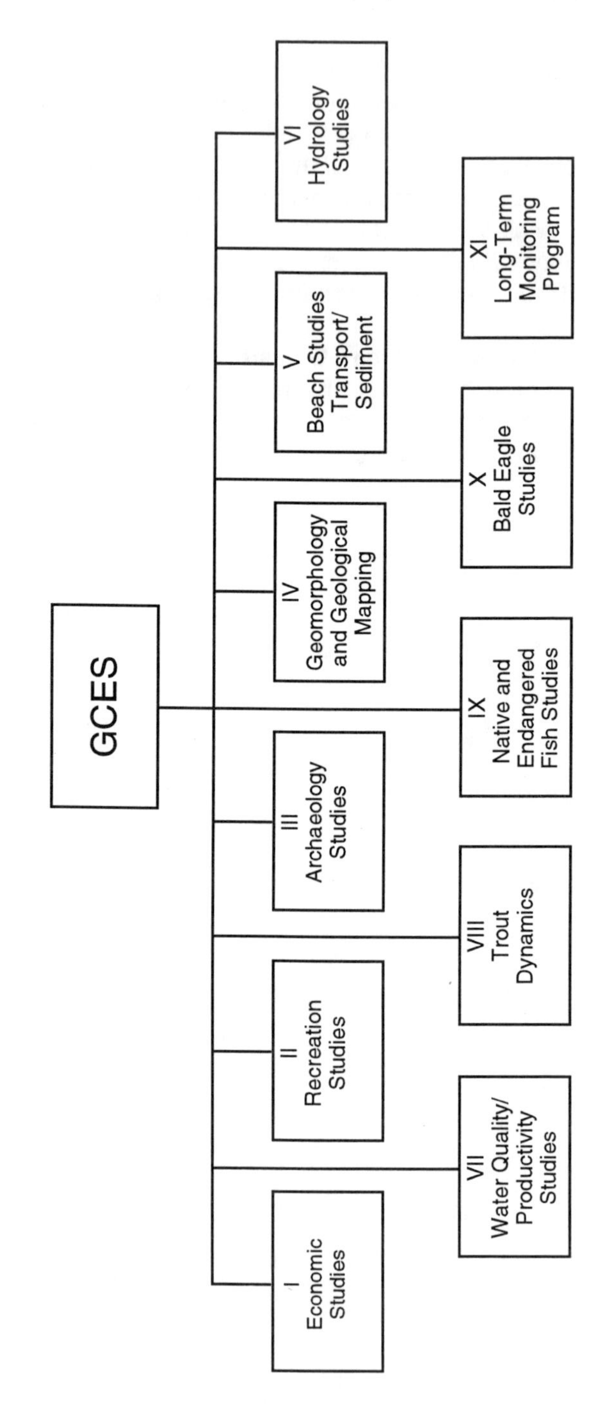

FIGURE 1.2 Components of GCES Phase II. SOURCE: Valdez and Ryel (1995).

REFERENCES

Andrews, E.D. 1991. Sediment transport in the Colorado River Basin. Pp. 54-74 in Colorado River Ecology and Dam Management. Washington, D.C.: National Academy Press.

Bureau of Reclamation. 1993. Operation of Glen Canyon Dam Colorado River Storage Project, Arizona. Draft Environmental Impact Statement, U.S. Department of the Interior, Washington, D.C.

Dawdy, D.R. 1991. Hydrology of Glen Canyon and the Grand Canyon. Pp. 40-53 in Colorado River Ecology and Dam Management. Washington, D.C.: National Academy Press.

Hughes, T.C. 1981. Reservoir operations. Pp. 207-225 in Colorado River Ecology and Dam Management. Washington, D.C.: National Academy Press.

Ingram, H., A.D. Tarlock, and C.R. Oggins. 1991. The law and politics of the operation of Glen Canyon Dam. Pp. 10-27 in Colorado River Ecology and Dam Management. Washington, D.C.: National Academy Press.

Martin, R. 1989. A Story that Stands Like a Dam: Glen Canyon and the Struggle for the Soul of the West. New York: Holt.

National Research Council. 1987. River and Dam Management: A Review of the Bureau of Reclamation's Glen Canyon Environmental Studies. Washington, D.C.: National Academy Press.

National Research Council. 1991. Colorado River Ecology and Dam Management. Washington, D.C.: National Academy Press.

National Research Council. 1994. Review of the Draft Environmental Impact Statement on Operation of Glen Canyon Dam. Washington, D.C.: National Academy Press.

Plummer, B. 1983. The Colorado River, a river for many people. Pp. 3-11 in Aquatic Resources Management of the Colorado River Ecosystem, D.D. Adams and V.A. Lamarra, eds. Ann Arbor, Mich.: Ann Arbor Science.

Powell, J.W. 1895. Canyons of the Colorado. Reprinted by Argosy-Antiquarian, New York, 1964.

Reisner, M. 1986. Cadillac Desert: The American West and Its Disappearing Water. New York: Viking Press.

Valdez and Ryel. 1995. Life history and ecology of the humpback chub (Gila cypha) in the Colorado River, Grand Canyon, Arizona. Final Report to the Bureau of Reclamation, Contract No. 0-CS-40-09110. Bio/West Report No. TR-250-08.

2

Scope and Organization of the Glen Canyon Environmental Studies

INTRODUCTION

The scope of the Glen Canyon Environmental Studies (GCES) of the Bureau of Reclamation program changed considerably between its inception in 1983 and its termination in 1995. In fact, one weakness of GCES was the instability of its conceptual and geographic boundaries. The scope and organization of GCES are instructive in part because they reflect beneficial maturation of the Bureau of Reclamation's (BOR) environmental analysis effort and in part because they illustrate some difficulties inherent in government environmental research projects.

SCOPE AS DEFINED BY MANAGEMENT OPTIONS, RESOURCES, AND THE ECOSYSTEM CONCEPT

Management Options

The objective of GCES was to establish a basis for forecasting the ways in which various options for managing Glen Canyon Dam would affect all resources of value to society. In light of this objective, it is remarkable that GCES management failed to develop and feature a comprehensive list of management options in planning GCES. Part of the explanation lies in the resistance of the BOR itself to the consideration of management options that were beyond its planning horizon. For example, the BOR, with encouragement from the U.S. Department of Energy's Western Area Power Administration (WAPA), initially resisted any notion that the dam could be managed in a way that would be inconsistent with the production of maximum power revenues as reflected by historical operation of Glen Canyon Dam (BOR, 1995). Thus, GCES at first was restricted to studies that reflected far less than a full slate of options. In fact, the resistance to consideration of all options extended well beyond BOR to virtually every other agency and constituency. Agencies, groups, and individuals, including scientists, have commonly judged management options on a mainly intuitive or single-factor basis.

As pointed out in the first National Research Council (NRC) review of GCES (NRC, 1987), GCES under the direction of the BOR should have been planned around a list of all management options not precluded by law or unrealistic in cost or feasibility. Table 2.1 provides such a list. As shown by the table there is considerable flexibility for management of resources through variations in the operation of Glen Canyon Dam within the legal requirements for delivery of water and prudent protection of the dam from catastrophic flooding. Had such a list originally been part of GCES planning, the studies would have been directed more quickly toward the questions that management ultimately faces. Instead, the items on the list were acknowledged incrementally over about a decade.

During preparation of the operations environmental impact statement (EIS) in 1994, and with encouragement from the 1992 Grand Canyon Protection Act, the BOR embraced the principle of adaptive management. In following this laudable principle, the BOR committed itself to frequent review of management options (Table 2.1), along with a list of resources, and to adjust management practices as necessary to optimize the aggregate value of all resources. GCES would have proceeded more efficiently had the BOR recognized this need initially rather than toward the end of the program, but the final adoption of adaptive management through the EIS is an important achievement for GCES and for the BOR.

Resources

Resources potentially affected by variations in the operation of Glen Canyon Dam are listed in Table 2.2. Until the BOR began responding to the results of GCES (ca. 1985), the dam had been managed entirely around water storage and delivery, flood prevention, and the production of power revenue. Water storage and delivery are not subject to administrative modification because they are fixed by law and because flood control is essential for

TABLE 2.1 Categories of Operational Flexibility Not Prohibited by Law

Type of Operational Flexibility	Probable Benefits	Probable Costs
I. Discharge		
A. Adjust seasonal discharge pattern	Improve aesthetic appeal, favor native species	Reduce power revenue
B. Adjust daily discharge pattern	Improve rafting aesthetically, stabilize backwater habitat for aquatic organisms	Reduce power revenue
C. Add controlled floods (ca. 50,000 cfs)	Lift sand to beaches, scour backwaters, move debris blockages, suppress exotic fishes	Reduce power revenue[b]
II. Temperature		
A. Warm river water by using multiple-level outlet withdrawal structure[a]	Encourage native species	Open migration to more exotic fishes
III. Turbidity, sediment supply		
A. Add sediment at dam by slurry pipeline[a]	Restore beaches in Glen Canyon	Suppress trout, high construction costs

[a]Not presently available; would require special EIS and construction authorization.
[b]Controlled floods may reduce power revenues negligibly if they can be timed to coincide with storm flows from the Little Colorado River.

TABLE 2.2 Resources Potentially Affected by Variations in the Operation of Glen Canyon Dam

I. Recreation
- A. Trout Fishing
- B. Rafting

II. Hydropower and hydropower revenues

III. Biotic communities
- A. Aquatic communities (including endangered species)
- B. Riparian communities (including endangered species)

IV. Sites of cultural or archaeological significance

V. Nonuse values

protection of the dam. Marketing of hydropower is required by law but is flexible operationally and thus confers on management the responsibility to assess the relationship between power production and the value or welfare of other resources. GCES was the first systematic attempt to describe and quantify the effect of dam operations on resources other than hydropower.

GCES did not initially make simultaneous commitments to studies of all the resources potentially affected by operations. It focused at first on sediment supply and sediment dynamics (see Chapter 5), which are important for recreation (Chapter 7) and for fishes, and on biotic communities, with particular emphasis on endangered species (Chapter 6). GCES was slow to acknowledge the relevance of cultural studies (Chapter 8) and initially excluded studies of power production (Chapter 9) because the BOR and WAPA began with the assumption that the dam would continue to be managed for production of maximum power revenues.

Even after the BOR accepted the modification of operating regimes for the benefit of other resources, with the loss of some power revenues, it resisted open (i.e., involving external independent analysis) studies of power production because studies of power had been internal to the BOR and WAPA since operation of the dam began. The EIS on power production did what GCES could not do initially, which was to require an open and complete study of power production in relation to dam operations. This was ultimately an extremely important contribution to the understanding of resources in relation to operations. Without it, GCES would have been far less valuable because it would not have led to reliable quantitative estimates of power revenue losses in relation to operational changes offering environmental benefits.

As shown by Chapter 9, studies of power and power revenue proved to be far more controversial and problematic than originally expected. Models in use by WAPA at the onset of the studies projected costs that were unjustifiably high and even outrageous for modest operational changes (NRC, 1991). The formation of a study group involving not only the BOR and WAPA but also environmentally oriented experts and independent consultants ultimately produced by consensus a very different view of power production revenues.

Another resource of interest that remained long unstudied is shown in Table 2.2 as nonuse values. As described in Chapter 7, nonuse value refers to that aspect of an environmental resource that derives from appreciation of a particular state of the resource by those who are not using it. Although relatively new, this is now an acknowledged dimension of comprehensive environmental studies (Chapter 7). Nonuse value seems particularly relevant in the case of the Grand Canyon because of the high aesthetic and intangible values attached to the region nationally and internationally and by Native American tribes (the BOR received 33,000 comments on the draft EIS for operation of Glen Canyon Dam). Even so, or perhaps for this very reason, the bureau long resisted inclusion of nonuse value studies but in 1995 acceded to them as an addendum to the EIS. Not surprisingly, the studies of nonuse value, which were conducted by an independent consultant with admirable use of peer review and outside critique, produced estimates of nonuse value that vastly exceed all tangible values, including power production revenues (Chapter 7). While the administrative response to this information is yet to be resolved, the information itself is clearly warranted as a component of GCES.

The Ecosystem Perspective

In 1987 the NRC committee was concerned about evidence that GCES lacked an appropriate conceptual framework around which to build its study and prioritize its expenditures. Phase II of GCES, however, embraced the ecosystem concept as a basis for planning. This advance was very important, but various factors prevented its full execution.

The ecosystem concept, which now has been widely embraced by federal agencies under the heading of ecosystem management (Lewis, 1994), is essentially a substitute for using a list of issues or resources in organizing environmental studies. While such lists continue to be important in all environmental studies, including those of the Colorado River, the study plan must take into account the relationships among physical, chemical, biotic, and anthropogenic components of the environment. The ecosystem concept acknowledges that management of environmental systems operates within a framework whose components are functionally connected.

Following NRC criticism (NRC, 1987) that the GCES study plan was haphazard and diffuse, GCES managers adopted an explicit ecosystem framework for the design and operation of studies. This move was reinforced by the appointment of a senior scientist, also in response to NRC recommendations. One purpose of appointing the senior scientist was to add expertise in ecosystem analysis to the GCES.

The most common way of applying the ecosystem concept to a particular location is by diagrammatic representation of the system. In the process of constructing such a diagram, analysts of the system must acknowledge in a comprehensive way the interactions that are likely to be critical to an understanding of the problems at hand.

The GCES produced its ecosystem diagram (Figure 2.1) when its program was already well under way. Critique of the GCES ecosystem diagram is possible, especially given the hindsight provided by a decade of GCES. On the whole, however, the diagram captures in a reasonable way the major components of the system and their relationships, especially with regard to the controversies that motivated GCES. By the time the diagram had been produced and refined, however, disposition of GCES resources was dictated either by precedent from earlier GCES studies or by legal or administrative fiat (see below). Thus, GCES did not provide an ideal test of utility for the ecosystem diagram.

As a test of the importance of the ecosystem diagram to GCES, in 1994 the NRC committee inquired about GCES support for each of the ecosystem components and causal connections shown in the diagram. This exercise demonstrated that a number of the significant causal connections shown in the diagram were not under study by GCES. In this sense, GCES was only incidentally an ecosystem study, even though it did ultimately cover quite a number of components and connections. Flaws in the diagram are brought out in the chapters of this report on individual components of the system. The primary flaw related to the ecosystem concept for GCES was, how-

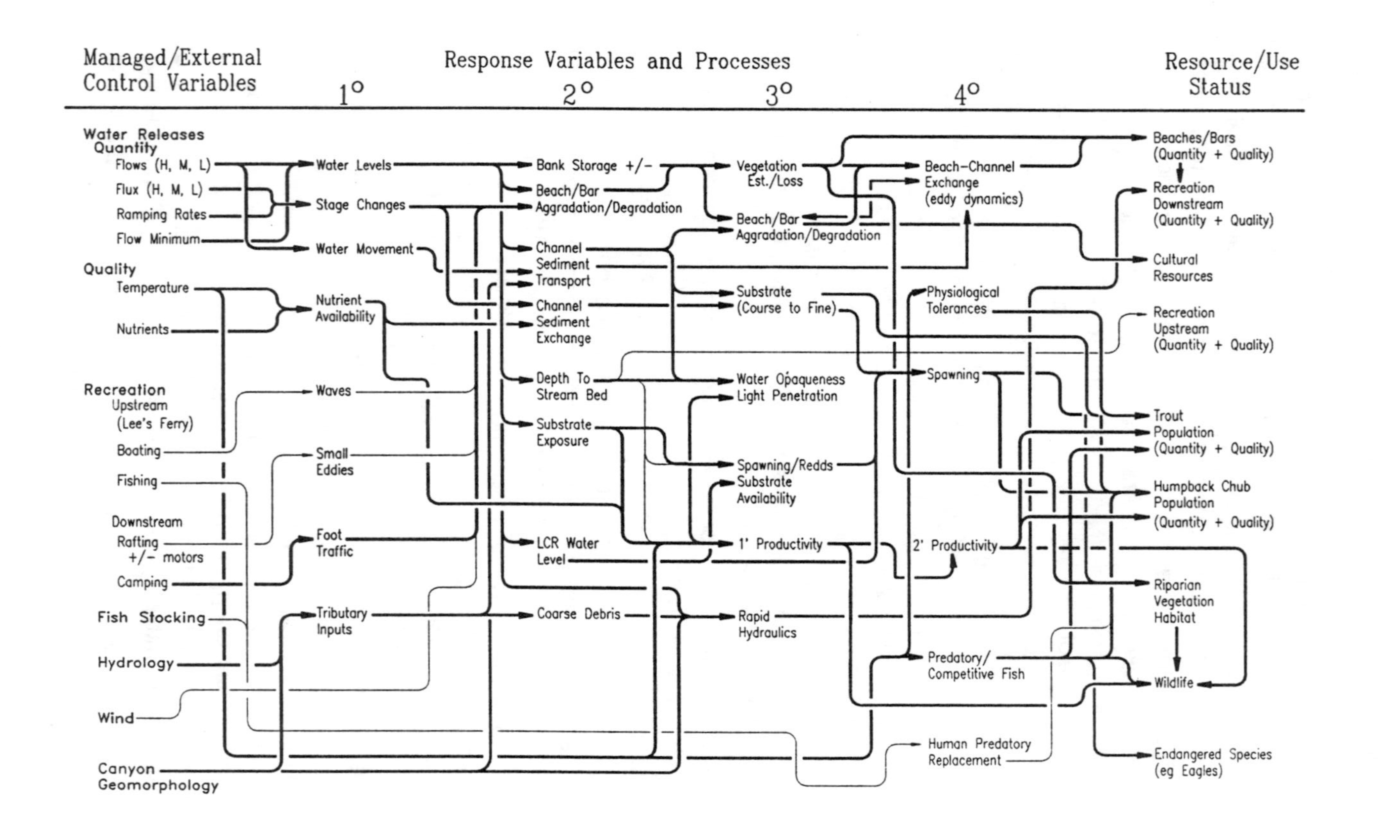

FIGURE 2.1 Ecosystem diagram used for GCES Phase II. Heavy lines show interactions of greatest importance.

ever, practical rather than conceptual.

The scope of work for GCES, and the priorities for GCES investments, were ultimately defined almost entirely by the list of resources shown in Table 2.2. The weakness of this approach is that each resource has specific advocates, which may be agencies, public groups, or commercial interests. Advocacy thus becomes a key factor influencing the design of studies, and the essential connection to operations is lost, as is the concept of ecosystem analysis. For example, GCES resources were used for studies of bald eagles, which began to populate the Grand Canyon seasonally in considerable numbers during the 1980s. There is at best a very weak connection, however, between the welfare of the bald eagle population and any conceivable variant on operations of the Glen Canyon Dam. Many examples, accounting ultimately for as much as half of the GCES budget, can be traced to this type of boundary expansion of GCES. The reasons for boundary expansion can be found mainly in forces associated with law and politics. The GCES was never completely driven by an ecosystem model. Rather, its focus changed over time as various external constituencies were able to convince Congress and the Department of the Interior to intervene in GCES and the EIS process.

OTHER INFLUENCES ON THE SCOPE OF GCES

Administrative Policy

The initial definition of scope for GCES was almost purely administrative in the sense that it did not reflect the geographic or conceptual extent of effects that might reasonably be connected with dam operations. The BOR directed GCES to deal with the Colorado River immediately below Glen Canyon Dam and to consider only changes in operations that would be neutral with respect to power revenues. One objective of the BOR was to contain the cost of GCES. Another factor, however, may have been the desire of the BOR and of WAPA with which BOR necessarily has a close working relationship, to reduce the likelihood of challenges to the principle that dam operations can be changed in ways that reduce power revenues, if environmental benefits accrue from such changes.

The initial administrative definition of scope was unrealistic. The flexibility of dam operations within the constraint of maximum power revenues is so small that studies restricted to this scope would be of limited use.

Expansion of GCES was probably inevitable given the range of interests and political forces that came to bear, especially through the EIS, on the BOR. The NRC committee played an important role in questioning the premise that flexibility of operations could be viewed only within the constraint of maximum power production. To its credit, the BOR ultimately accepted the advisability of viewing the scope of GCES much more broadly and thus more realistically. By the end of GCES, the initial unrealistic constraints on its scope had disappeared.

In its last several years GCES came under the influence of factors that were not always directly related to the effects of dam operations. GCES expanded geographically to the upper end of Lake Powell, up the tributary canyons, above the present high-water mark, and out to tribal lands on the rim. A rational administrative approach would have been to ask, as a means of limiting scope, whether a given type of study or inquiry could reasonably be connected to variations in the operation of Glen Canyon Dam. Instead, the BOR adopted the more expedient principle of simply setting an annual dollar cap (often a quite generous one, extending to as much as $12 million per year at its peak) on the GCES and essentially ignoring the necessity of connections between scope and the dam's operation. As explained below, several factors contributed to this change in administrative policy.

Law and Politics

In several instances the scope of GCES was affected by federal laws that extended the program well beyond the connections between resources and dam operations. When GCES began, no EIS was in progress or was planned. When Interior Secretary Lujan invoked the National Environmental Policy Act (NEPA) by calling for two EISs (one for operations and one for power production), the scope of GCES was affected. The NEPA checklist for EIS production called repeatedly for information that could only come from GCES and thus modified GCES scope and scheduling. For example, studies of archaeological sites by the National Park Service were arguably far more elaborate than they needed to be if defined by the original objectives of GCES rather than the more inclusive requirements of the EIS, which were augmented by the Park Service's interest in obtaining an archaeological inventory with monies originating outside its budget.

As explained more fully in Chapter 10, individual agencies used their mission statements as guiding principles in defining research objectives rather than the specific needs of GCES. The research conducted by a myriad of cooperators under the GCES umbrella would have been more effective if all the parties involved had devised a system to focus on resources on the stated purpose of GCES. Much of the information that was collected outside the obvious scope of GCES is scientifically sound and will prove useful in other contexts, but the inability of the "cooperators" (Table 2.3) to devise a system for focusing resources on the stated purpose of GCES was a remarkable and consistent feature of the program, and resulted in great expansion of expenditures and diffusion of focus. Because federal laws and high-level politics were involved in these processes, it is impossible to fault individuals and certainly not the primary organizers of GCES, who to a large extent were buffeted by forces well beyond their control.

In 1987 the NRC committee criticized the GCES for failing to derive and follow a specific plan that, in turn, would be linked to alternative possibilities for operating Glen Canyon Dam. While the committee was quite successful in encouraging a badly needed expansion in the scope of GCES during the mid-1980s, it was later much less successful in trying to move GCES toward greater focus of resources on a predetermined plan and exclusion of expenditures that could not be justified by the objectives of the program.

ORGANIZATION OF THE STUDY GROUP

By the early 1980s, when GCES began, patterns for the use of power revenue from Glen Canyon Dam had been in place for almost 20 years. The intrusion of GCES as a new demand on the power revenue stream was vehemently and openly opposed by WAPA, which argued that power revenues ($50 million to $100 million per year, depending on the market for electricity) should not be used for this purpose. Nevertheless, roughly 10 percent of power revenues was used for GCES over its 13-year life. As a result of the Grand Canyon Protection Act of 1992 (see Chapter 3), long-term monitoring, which is the successor to GCES, will constitute an expense that is reimbursable to the U.S. Treasury after the EIS Record of Decision is issued. Thus, future costs of environmental studies below Glen Canyon Dam will be borne by U.S. taxpayers at large, rather than by users of power from Glen Canyon Dam.

Although the GCES was initiated and managed by the BOR, it involved extensive participation by other federal and state agencies and by Native American tribes. Decisions about the course of GCES were made by the BOR manager but typically with strong advice from the cooperators. The

TABLE 2.3 Organizational Participants in GCES

Cooperators	GCES Expenditures 1983-1994	Areas of Particular Interest
Federal agency		
1. U.S. Bureau of Reclamation	$1,945,900	Management, hydropower modeling
2. U.S. Fish and Wildlife Service	$2,001,198	Endangered fish
3. U.S. Geological Survey	$14,061,605	Sediment, hydrology
4. U.S. National Park Service	$14,168,301	Recreation, archeology, biology, sediment
State agency		
5. Arizona Game and Fish Department	$5,483,466	Trout, native fish, aquatic food base
Tribal		
6. Havasupai Tribe	$50,000	Hydrology
7. Hopi Tribe	$2,012,529	Hydrology, archeology
8. Hualapai Nation	$2,257,411	Archeology, recreation, biology
9. Navajo Nation	$3,435,695	Archeology, biology
10. Paiute Tribe	$320,000	Cultural
11. Zuni Tribe	$505,804	Cultural
Major Contractors (not cooperators)		
12. BioWest	$3,906,052	Endangered fish
13. Northern Arizona University	$78,000	Trout
14. University of Arizona	$130,942	Sediment
15. Arizona State University	$1,378,194	Endangered fish
16. National Research Council	$1,237,338	Review of GCES

cooperators not only provided advice on the course of study but also received and expended GCES monies or served as conduits for GCES contracts (Table 2.3).

The organizational structure of GCES contained several weaknesses that became increasingly evident over its history. While the BOR should be credited with assigning a full-time position to the management of GCES and placing the GCES budget in principle at the disposal of the project manager, the BOR failed to reinforce the independence and effectiveness of project management. The immediate reporting hierarchy of the project manager was initially critical of the fundamental basis for GCES and particularly of its expansion to a realistic geographic and conceptual scope. Although this situation changed during the last years of GCES, it handicapped the early phases of project development. Equally important was the level of administrative placement of the GCES management, which reported to a district office of the BOR.

The BOR is an agency of the U.S. Department of Interior, as are the cooperators, with the exception of the Native American tribes and the Arizona Game and Fish Department. Thus, the GCES project manager, who was reporting to a district administrator inside one of the agencies among the cooperators, was outranked by most of the individuals who made up the cooperating group. The NRC committee perceived this difficulty early in GCES and recommended in 1987 (NRC, 1987) that GCES Phase II be organized in such a way that the project manager would report to an Assistant Secretary of the Interior, thus being connected to the administrative umbrella over all of the federal cooperators. This change was not made.

The effect of administrative flaws on GCES is difficult to evaluate. Subsequent chapters of this report will show, however, that GCES suffered, notwithstanding energetic and adaptable leadership, from a consistent inability to exercise control over the cooperators for the benefit of general project objectives. In practical terms the project manager was unable to withhold funds from agencies that failed to meet contractual obligations and had difficulty in confining or directing the scope along lines that were contrary to those preferred by individual cooperators.

CONCLUSIONS

The initial design of GCES was severely flawed in definition of scope and in organization for a variety of reasons. Over several years, through adjustments made primarily through the response of the GCES management and the BOR to outside comment and critique, many of the flaws were reduced or eliminated. Thus, the scope and organizational strength of GCES improved over the life of the project. If these institutional adjustments can be carried into other projects of the BOR or other management agencies, future environmental projects will be less costly and more effective in meeting their goals.

While GCES improved in a number of important respects over its history, its effectiveness was impaired even in its latest phases by a number of unsolved problems related to scope and organization. These include especially the internalization of expenditures among the federal cooperators, the inability of the project manager to exercise free authority over critical decisions because of the administrative structure of the project, and the introduction of ancillary objectives through law or administrative fiat that were not necessarily relevant to the project's objectives.

The GCES has shown that federally sponsored environmental assessment should be organized around three sets of considerations: 1) a list of resources, 2) a list of management options, and 3) the ecosystem concept. Omission or incomplete treatment of any of these three considerations will greatly impair the usefulness of the final outcome of environmental assessment. Furthermore, given that large, federal environmental studies will in the future increasingly involve multiple federal agencies with differing missions and priorities, the project manager for any large environmental assessment must be granted, for the benefit of the project, sufficient independence and authority over financial resources to override undue influence by individual agencies. The GCES experience shows that concentration of authority in the project leadership, and initial commitment to complete consideration of all management options and resources, including those that may be out of favor or controversial, will be the most likely strategies to conserve resources and produce outcomes useful to management.

REFERENCES

Bureau of Reclamation. 1995. Operation of Glen Canyon Dam. Final Environmental Impact Statement. U.S. Department of the Interior, Washington, D.C.

Lewis, W.M., Jr. 1994. The ecological sciences and the public domain. University of Colorado Law Review 65:279-292.

National Research Council. 1987. River and Dam Management: A Review of the Bureau of Reclamation's Glen Canyon Environmental Studies. Washington, D.C.: National Academy Press.

National Research Council. 1991. Colorado River Ecology and Dam Management. Washington, D.C.: National Academy Press.

3

Historical Context for Long-Term Management of Glen Canyon Dam

The federal government agencies with a stake in the operation of Glen Canyon Dam and the Colorado River states have concentrated on two related questions: How should the dam be managed, and how should the impact of its operations be monitored? These questions cannot be answered, however, until a third question is raised and answered: For what objectives should the dam be managed? This question was never clearly addressed by the Glen Canyon Environmental Studies (GCES) or by the dam's managers or other interested parties. The environmental impact statement (EIS) on the operation of Glen Canyon Dam (BOR, 1995) suggests that the dam should operate in a way that minimizes adverse effects on endangered species, recreation, and cultural resources to the extent that this can be done without substantially modifying the traditional operating priorities for the dam.

The Law of the River is the legal accretion of judicial, legislative, and compact resolutions of historic conflicts among diverse user groups (NRC, 1991). Mitigation is a logical counterpart to the Law of the River, but it is too narrow a perspective for the management of the Colorado River through the Grand Canyon. The expanding demand for use of the river has produced numerous groups of well-defined and well-organized stakeholders. As trustees for their citizens, the Colorado River basin states have asserted their claims to a share of the river, and individual groups of water users such as irrigation districts, utilities, and municipal water suppliers have obtained water rights to the river and contract rights to the power generated by the dams on the river. The Colorado River has long been allocated among the seven basin states by interstate compacts, congressional legislation, and Supreme Court decrees. Each basin state has a share of the river in perpetuity to distribute to users in the state. The lower-basin states have used their full entitlements to support irrigated agriculture and unlimited urban growth; the upper-basin states are trying to do the same but have had less demand. The accommodation of traditional stakeholders, however, is no longer the sole function of dam managers. There are two related reasons for this.

The federal dam managers were initially "trustees" for the basin states and individual states holding water rights, but their responsibilities have expanded over time. The large dams were built to provide a reliable source of water and to finance, through power revenues, costly distribution systems that states, cities, and irrigation districts could not afford. The federal role has been to subsidize regional water development. When the water was allocated and the large reservoirs were complete, the federal interest declined. Aside from the possibility of federal reserved water rights for Indian tribes, the federal interest is now confined to the recoupment of the monies spent on the dams and irrigation projects and to management of existing facilities. When the Bureau of Reclamation (BOR) first began the GCES, it assumed, as did the Western Area Power Administration (WAPA), that the federal role was unchanged from the 1920s and 1930s. The GCES, however, revealed that the list of stakeholders has expanded beyond the states and individual holders of water rights and that federal agencies have management duties that do not derive from the Law of the River as it has been historically understood.

The new claimants include Indian tribes, which are asserting quasi-sovereignty over portions of the river and its associated environment, recreational users such as river rafters and sport fishermen, and diverse environmental groups. Indian tribes have asserted water rights and now assert broader claims to protect sacred or religious sites in the river corridor and to participate in the management of the dam. River rafters have permit entitlements to run the river. Environmental groups have asserted a wide variety of interests from haze reduction over the Grand Canyon (NRC, 1990) to modification of dam releases to protect endangered species in the canyon.

Mitigation of identified adverse environmental impacts is the legal strategy that the nation has followed since the passage of the National Environmental Policy Act of 1969. Agencies must use the EIS process to identify adverse environmental impacts and then are expected to mitigate them unless there are strong reasons, in the context of preexisting programmatic mandates, why they should not. Mitigation as a long-term strategy is limited because it is purely reactive. This strategy is supported by the assumption, grounded in pluralist democratic theory, that the dam should be managed to accommodate the interests of all of the major stakeholders on the river as reflected in their actual or potential legal entitlements.

The GCES can be praised for opening the scientific evaluation of dam operations to a wide variety of stakeholders. Traditional river user groups and the national environmental community have been sufficiently well organized and funded to participate in review and evaluation of GCES studies. GCES has also served other groups; the involvement of Indian tribes is particularly significant. The focus on present user groups, however, ignores the intergenerational dimension of Grand Canyon management. Focus on the mitigation of selected adverse impacts does not always produce the most effective science (see Chapter 2). Scientific research organized in this way runs the risk of being fragmented; there is little incentive to integrate the research into a broader framework. In addition, better management does not necessarily follow from the results of investigation that is overly directed by a list of specific issues.

Of all the stakeholders, the National Park Service (NPS) would be the most likely candidate to formulate management objectives for the canyon. This did not happen. The history of Grand Canyon National Park provided no basis for the development of a management perspective because the river corridor has never been the focus of the NPS's mission. The primary interest of the NPS in the river corridor has centered around issuing and monitoring float trip permits.

The NRC committee's 10-year experience with the GCES suggests that the concept of ecosystem management is a better management model than mitigation (Chapter 2). Mitigation is an important component of ecosystem management, but it is neither the starting nor the ending point. Ecosystem management does not attempt to fix discrete adverse effects of an activity, but rather to maintain the vital functions of a natural system as modified by human activity over time, through adaptive management. The basic idea is to develop background norms and then use them to measure the health of the system. This does not mean, as it is sometimes understood, that the objective of management is to return the system to a totally natural condition. The system can be managed for the optimization of any mix of objectives, but the ecosystem perspectives recognize the inevitable connection of any management scheme to all resources. As GCES evolved, the need to place the operation of Glen Canyon Dam in the broader context of management of the river corridor through the park and adjoining national recreation area became clearer. The next section of this chapter discusses the history of Grand Canyon National Park. Its purpose is to show that although the Colorado River has been central to the formation of the unique geology that constitutes the park, it has never been central to the park's management mission. This may be the root of one of the central problems in GCES: the discontinuity between the functions of the Colorado River and the agencies that control the flow of the river.

HISTORY OF GRAND CANYON NATIONAL PARK

Introduction

The traditional focus of the National Park Service (NPS) in the Grand Canyon has been the protection of scenic vistas rather than preservation of the ecological integrity of the Colorado River corridor. The motivation for GCES and for the EIS on dam operations is that Glen Canyon Dam alters the flow of the Colorado River through a world-renowned national park, but this has not been a traditional concern of the most likely stakeholder, the National Park Service. Not only is the Grand Canyon a unit of the NPS, it is a world heritage site pursuant to the Convention on World Heritage Sites. An outsider looking at the history of the GCES, however, would be surprised at the minimal role played by the NPS in defining the objectives and scope of the program (Babbitt, 1990).

Predesignation History

Grand Canyon National Park is one of the crown jewels of our country's park system. John Wesley Powell's 1869 voyage through the canyon brought the scenic wonders of the area to public attention in the East. Legislation to designate the canyon as a national park was introduced in 1882, but Congress did not act on it for over 35 years. The reasons lie in the politics of the late frontier and in the lack of access to the canyon. There was no easy access until the Santa Fe Railroad built a spur from its mainline across northern Arizona in 1901 and constructed the El Tovar Hotel in 1904. The area was designated as a national park in 1919, following its designation as a national forest in 1893, a game reserve in 1906, and a national monument in 1908 (Ise, 1961).

Designation as a Park

After the Santa Fe Railroad began to develop the Grand Canyon, legislation calling for preservation of its natural resources was introduced between 1905 and 1919 but was not enacted until vested rights claims were settled. Subordination of the park to nonpark interests continues today. The Grand Canyon was designated as a park 3 years after the NPS was established, and its management has been strongly influenced by former NPS Superintendent Mather's administration philosophy, which is to encourage visitor access to national parks. The park was originally opposed by the Arizona business community because it would preclude mining and the development of private concessions. However, as visitor use increased, the economic value of the park became apparent to the business community, which then supported its designation as a park. A powerful political figure and later U.S. Senator, Ralph Henry Cameron, had, however, located a mining claim in the park in 1908, and preservation of existing valid claims was one of the conditions for the park's enabling legislation.

Much of the NPS's early efforts were devoted to elimination of the Cameron mining claims. Henry Cameron tried to control the tourist business at the rim by locating 45 mining claims on the South Rim at the head of Bright Angel Trail in 1908. He was able to control access to the trail and Indian Gardens "to the distress of the Forest Service, the Santa Fe Railroad and, tourists" (Ise, 1961). His claims were ultimately invalidated by the Supreme Court, but Cameron was elected to the Senate in 1920. From his Senate seat, he was able to harass the NPS and tried to regain control of the trail by enacting a rider in the NPS's appropriations act forbidding the use of federal funds for trail maintenance. His influence waned after the Teapot Dome scandal, but mining was not outlawed until 1931.

The Grand Canyon was set aside as a national park for aesthetic reasons, and thus the rim rather than the canyon corridor was the object of early NPS preservation efforts (Leydet, 1964). As J. Ise, leading historian of the national parks wrote, the "Grand Canyon is remarkable mainly as our most spectacular scenic wonder" (Ise, 1961). The national parks were established in a spirit of cultural nationalism; they were a substitute for the man-made monuments of Europe (Runte, 1979). As a result, the rim view was treated as the primary attribute of the canyon. The more adventurous visitors took Bright Angel Trail to Phantom Ranch, but the river was not central to these activities. Viewed from the rim, the huge hydrological variations of the unimpounded river were barely perceptible.

In U.S. law regarding public lands, national parks have single rather than multiple purposes, and the preservation of scenic grandeur is the historic statutory mission of the NPS. Section 1 of the NPS 's enabling legislation (16 U.S.C. § 1) requires the Secretary of the Interior to manage the system units "by such means as will leave them unimpaired for future generations." In 1878, Congress supplemented this mandate by directing the Interior Secretary to administer the parks "in light of the high public value and integrity of the National Park system" (16 U.S.C. § 1a-1). No specific additional mandates are found in the legislation governing the Grand Canyon. The general mandates are, however, misleading. Parks are federal land management units subject to competing demands, and the actual mission of the NPS has been the promotion and accommodation of visitor use.

The history of Grand Canyon National Park is one of constant expansion of visitor comfort and services and a consequent decrease in visitor appreciation of the canyon as wilderness. This is the legacy of Stephen Mather and Horace Alright. These first two park superintendents established a powerful constituency for the idea of national parks by promoting easy access and visitor enjoyment (Forestra, 1983). Because the main canyon asset has been perceived as primarily geological, there has been insufficient attention to other components of the ecosystem (Nash, 1983). Until the 1960s, boat trips down the canyon were limited to scientific surveys or a few intrepid adventurers, some of whom lost their lives (Hughes, 1978). In fact the NPS has historically taken the position that the river corridor does not qualify for designation under the Wilderness Act of 1964 because of the historic and continuing use of outboard motors on it.

The importance of the view at the Grand Canyon is illustrated by the federal government's responses to haze. In 1968, legislation to complete the Colorado River storage plan with a reservoir at either end of the canyon, and to finance construction of the Central Arizona Project with revenues from these, was defeated after a national political campaign led by the Sierra Club. Ironically, as a substitute for the defeated hydro projects, a coal-fired power plant, the Navajo generating station, was built near the dam in Page, Arizona. The plant contributed greatly to a persistent haze in the Four Corners Area (NRC, 1990), and in 1990 Congress attempted to resolve the issue by requiring the Navajo generating plant to install scrubbers.

Role of the Colorado River in the Park: Conduit Between Upper and Lower Basins

In retrospect, it is amazing that the Colorado River corridor is as undeveloped as it is today. In 1919 the river had not been allocated between basins, but the necessity for an allocation and for carryover storage reservoirs to support the allocation were recognized by western politicians, and the pending allocation of the Colorado River influenced the enabling legislation. The enabling legislation (16 U.S.C. § 227) permits use of the canyon for reclamation projects and authorizes construction of reclamation projects.

Historically, the river's primary function has been as a conduit between the upper and lower basins. The seven Colorado River basin states have been given mass allocations by interstate compacts, congressional legislation, and Supreme Court degree. In addition, the claims of Mexico have been recognized by treaty and executive agreements. Under the 1922 Colorado River Compact, which allocates the river between the upper and the lower basins, the river is divided at Lee's Ferry above the canyon. Each basin was allocated 7.5 million acre-feet (maf), because 15 maf was erroneously assumed to be the average annual flow of the river. Each basin shares equally in the obligation to provide an additional 1.5 maf to Mexico. To allow the more slowly developing upper basin to meet its delivery obligations to the lower basin, the compact defines the upper basin's delivery obligations to Arizona, California, and Nevada as 75 maf over a progressive series of 10-year periods, and the federal government has constructed two large carryover storage reservoirs to guarantee the upper basin's ability to meet this obligation during sustained droughts. Water moves through the river from the upper basin's storage reservoir behind Glen Canyon Dam to the lower basin at Boulder Dam in order to meet the upper basin's 8.3 maf annual compact and treaty delivery obligations to Arizona, California, Nevada, and the Republic of Mexico. The net result of the construction of these two storage and hydroelectric generating dams is that the Colorado River has become entirely regulated hydrologically (Fradkin, 1981). The Park Service's long history of trying to preserve the natural environment of the Grand Canyon rim (with the exception of visitor access facilities) makes it ill-equipped to manage regulated systems such as the Colorado River (Carothers and Brown, 1991).

Congress responded directly to new constituencies in the passage of the 1992 Grand Canyon Protection Act, although passage of the 1968 Colorado River Storage Project Act (43 U.S.C. §§ 620-620o) marked a fundamental change in the role of the river's corridor. The 1968 act reflected the defeat of efforts to construct two dams at either end of the canyon to finance the Central Arizona Project and authorized the operation of the dam for environmental protection as well as for power carryover storage and power generation (Marion and Wallick, 1991). The Grand Canyon Protection Act of 1992 is a direct outcome of the GCES's identification of the need for a different release pattern from the dam to both build beaches and retard beach erosion. In 1990, GCES scientists proposed a research flow program to test the impacts of less fluctuating flows and the spring beach-building pulses on the corridor. Legislation authorizing interim flows was introduced in the House and Senate in 1990 (H.R. 4498, 101 Cong., 1st sess., 1990), but the act was not passed for 2 years. Initially, the Department of the Interior opposed the legislation because the research flows had not been implemented and evaluated, but this opposition ended after BOR and WAPA agreed to an experimental interim flow regime in late 1991. The basic purpose of the act changed from a congressional mandate to a more general effort to expand management objectives. The act establishes the legality of river corridor enhancement flows consistent with the Law of the River.

Section 1802 of the 1992 act requires that the Secretary of the Interior operate the dam in a manner consistent with the Law of the River, including the Endangered Species Act "to mitigate adverse impacts to and improve the values for which the Grand Canyon National Park and the Glen Canyon National Recreation Area were established, including but not limited to natural and cultural resources and visitor use." The act makes the EIS the basis for future management. The 1991 agreement is continued, with limited exceptions, pending completion and implementation of the EIS (§ 1803). Section 1804 requires that the Secretary of the Interior use the "findings, conclusions, and recommendations" of the EIS to adopt management criteria and operating plans in addition to those specified in Section 602 of the Colorado Basin Project Act of 1968.

FUTURE BASIS FOR MANAGEMENT OF GRAND CANYON RESOURCES

Intergenerational Equity

The monies expended on the GCES can best be justified because society cares about the future of the river corridor and has some sense of obligation to future generations. Specifically, it is now widely recognized that the underlying philosophical principle of much environmental management is intergenerational equity. Under emerging norms of international environmental law, the United States holds the canyon in trust for future generations. The idea of intergenerational equity was developed by Weiss (1989) and has rapidly been adopted as the ethical norm against which major international agreements and mandates must be tested. The basic idea is that "[w]e as a species, hold the natural and cultural environment of our planet, both with members of the present generation and with other generations, past and future." The precise contours of intergenerational duties are not self-defining, but the core idea is that each generation has a duty to manage its common patrimony for the benefit of the next generation.

Adoption of intergenerational equity fundamentally changes the nature of the decision-making process regardless of the precise content of the duty. Present actions should rather be evaluated in terms of the long-term consequences, and all present-value economic calculations of commodity values should be weighed against calculations that estimate the future value of resources and incorporate the assumption that environmental quality is the marginal value of natural or nondegraded resources and is likely to increase over time. This is the essence of the difference between the economics of sustainable development and traditional cost-benefit calculations (Pearce et al., 1990).

Intergenerational equity has long been part of the Law of the River and of the GCES, but there was no systematic effort to articulate the principles and to apply them to GCES research and Glen Canyon management options. The 1922 Colorado River Compact apportions the river between the upper and the lower basins in perpetuity. Thus, the rights of future generations to a sustainable use of the resource are explicitly recognized. The GCES included studies of nonuse value (Chapter 7). Such studies implicitly reject the prevailing economic theory that present consumption is preferred to future consumption, except over short time horizons. The controversy over the legitimacy of nonuse values per se, as well as the techniques required to quantify them (such as contingent valuation) led the BOR and WAPA to oppose their use during the early stages of the GCES.

Simulated Naturalness

As indicated by the review of Grand Canyon history, the policies and objectives that apply to national parks have been largely disconnected from the management of Glen Canyon Dam, even though the dam influences the characteristics of the river and riparian environment through Grand Canyon National Park. Through GCES and through the EIS, the BOR has acknowledged that there is considerable latitude for variation in the operation of Glen Canyon Dam, even within the binding legal constraints for water delivery and the need to protect the dam from damage by flood. If there were no flexibility in operation, the effects of the dam on the national park downstream would simply be one of the baseline conditions for the national park. Given that operation is flexible, however, its management should take into account the presence of the national park downstream.

The present strategy of BOR and its cooperating agencies, as shown by the EIS on Glen Canyon Dam operations, is to acknowledge the potential effect of dam operations on a wide variety of resources, and to consider patterns of operation that reflect at least some concern for the welfare of all resources. This type of optimization strategy is a common principle for modern environmental management, and offers many beneficial possibilities for the Colorado River in the Grand Canyon. The principle of optimization does not, however, provide any firm objective for management because optimization is achieved by the assignment of differential weightings to various resources. These weightings are subject to large degrees of negotiation and professional judgment.

A different kind of management principle, which might be called the principle of naturalness, applies to national parks. Management is minimized, and where it must occur, it is directed toward the maintenance of environmental regime that as nearly as possible resembles the natural or undisturbed condition of the environment. It seems unreasonable to consider the future operation of Glen Canyon Dam without also considering the principle of naturalness as it might apply to the Grand Canyon National Park.

While many aspects of the Grand Canyon are in fact natural or at least not subject to management or direct human perturbation, the river itself and the riparian corridor inevitably are a reflection of human action because of the existence of Glen Canyon Dam. The dam will continue to exist and will inevitably be a means by which the downstream environment is managed, either haphazardly or toward particular goals. The GCES has shown that operation of the dam can be modified in various ways to restore a greater degree of naturalness to the river and riparian environments through maintenance or restoration of physical characteristics of the environment such as beaches or biotic resources such as endangered species. Given the emphasis of national parks on naturalness, and the flexibility of operations to restore some aspects of naturalness, one obvious basis for future management of Glen Canyon Dam might be characterized as simulated naturalness, which could be defined as the use of operational flexibility to restore and maintain environmental conditions in the national park that resemble as nearly as possible the original condition of the river.

Many aspects of the river corridor in Grand Canyon National Park cannot feasibly resemble the original river corridor. As shown by the chapters to follow, however, there are many ways in which the environmental conditions along the river can be restored to a more natural state. These possibilities, some of which are in place or under construction, include adaptation of a more natural hydrologic regime, the introduction of controlled floods, restoration of seasonally warm water in the river, and maintenance of habitat and physical features such as beaches through manipulation of water and sediment. The adoption of simulated naturalness would give a unifying theme and purpose to operational changes with these objectives, and would provide a blueprint for the future.

For many intensively managed environmental systems, including the tailwaters and pools of most reservoirs in the United States, it makes little sense to manage toward simulation of natural conditions. Instead, the more pragmatic optimization approach provides a tool by which societal preferences, tangible resource values, and operational flexibility can be brought together in a management plan. The Colorado River in the Grand Canyon is not, however, a typical tailwater. It lies in a national park and is thus subject to the special purposes that apply uniquely to national parks. One of these purposes clearly is the maintenance and restoration of conditions that are, within reasonable limits of human effort and expense, natural. Thus there is much to recommend the principle of simulated naturalness as a future basis for management of Glen Canyon Dam, even though such a principle might be unjustifiably confining for the operation of large reservoirs in general.

REFERENCES

Babbitt, B. 1990. Introduction: down the imperiled Colorado. Land and Water Law Review 25:1.

Bureau of Reclamation. 1995. Operation of Glen Canyon Dam. Final Environmental Impact Statement, March, U.S. Department of the Interior, Washington, D.C.

Carothers, S.W., and B.T. Brown. 1991. The Colorado River Through Grand Canyon. Tucson: University of Arizona Press.

Foresta, R. 1983. America's National Parks and Their Keepers. Washington, D.C.: Resources for the Future.

Fradkin, P. 1981. A River No More. New York: Alfred A. Knopf.

Hughes, J.D. 1978. In the House of Stone and Light: A Human History of the Grand Canyon. Arizona: Grand Canyon Natural History Association.

Ise, J. 1961. Our National Park Policy: A Critical History. Baltimore: Johns Hopkins University Press.

Leydet, F. 1964. Time and the River Flowing: Grand Canyon. San Fransisco: Sierra Club Books.

Marion, K., and D. Wallick. 1991. Glen Canyon Dam Operating Authority: Producing Electricity and Protecting the Grand Canyon Environment. Land and Water Law Review 26:183.

Nash, R. 1983. Wilderness Values and the Colorado River. Pp. 201-214 in New Courses for the Colorado River: Major Issues for the Next Century. G. Weatherford and F. Lee Brown, eds. Albuquerque: University of New Mexico Press.

National Research Council. 1990. Haze in the Grand Canyon. Washington, D.C.: National Academy Press.

National Research Council. 1991. Colorado River Ecology and Dam Management. Washington, D.C: National Academy Press.

Pearce, D., et al. 1990. Sustainable Development: Economics and the Environment in the Third World. Brookfield, VT: E. Elgar Publishing Co.

Runte, A. 1979. National Parks: The American Experience. Lincoln: University of Nebraska Press.

Weiss, E.B. 1989. In Fairness to Future Generations: International Law, Common Patrimony and Intergenerational Equity 17.

4

Operation of Glen Canyon Dam

INTRODUCTION

Glen Canyon Dam and Lake Powell represent a major source of water, energy, and flood protection in the southwestern United States. It is therefore not surprising that objections from groups that use water and energy appeared quickly when significant changes in the dam's operating rules were proposed for environmental reasons. While objections related to increases in the cost of energy did have a valid basis, concerns about changes in water supply, and flood protection did not. The nature of the changes resulting from the effort to protect the environment below the dam had no effect on the ability of upper-basin states to deliver lower-basin water (the Law of the River) or the degree of flood protection provided by the dam. The reason is because the monthly release targets are totally independent of any constraint related to either interim release rules or preferred alternative criteria of the environmental impact statement (EIS). These changes, which were introduced for purposes of environmental protection, speak only to variations during a day, not the average volume released during a day (and therefore the month or year). This conclusion is supported by the simulation model results included in the Bureau of Reclamation's (BOR) EIS. The median annual release is the same (8.5 million acre feet (maf)) for all of the operational rule alternatives considered (BOR, 1994b, Table II-7).

This chapter compares the hydrology of the Colorado River before and after construction of the Glen Canyon Dam, describes operating rules for the dam and their evolution, and provides an overview of water supply above the dam. These three topics define the scope within which the dam's operation can be adapted to environmental objectives.

HYDROLOGY THROUGH THE GRAND CANYON

Hydrology Prior to Construction of Glen Canyon Dam

Prior to construction of Glen Canyon Dam, the flows at Lee's Ferry varied seasonally (Figure 4.1). The average annual discharge of 16,800 cubic feet per second (cfs) included periods of high flows sometimes exceeding 100,000 cfs (June) and flows as low as 2,500 cfs (fall and winter). Monthly average (June) flows were much more constant following construction of the dam. This is not surprising given that a principal justification of the reservoir was to reduce spring floods and to store this water, thereby providing subsequent increases in low-flow seasons.

The daily predam fluctuations were much smaller than seasonal variations, but occasional daily fluctuations were very significant. For example, variations in stage height of 5 to 10 feet for 1 to about 5 days are shown in Figure 4.2. These variations occurred during all seasons because of precipitation in tributary watersheds or temperature variations during the snowmelt season. Figure 4.2 also shows that minor daily fluctuations of about 1 to 3 inches were very common.

Hydrology Following Closure of the Dam

Following the initial filling of Lake Powell in 1980, and until interim flows began in 1991, the average annual flows were unchanged except in response to short-term droughts or floods, but daily fluctuations in discharge were large, as shown by Figure 4.3. Daily operating rules reflected variations in peak demand for hydropower. Maximum controlled daily peaks approached the 31,500-cfs capacity of hydropower turbines, and daily minima were as low as 1,000 cfs in winter and 3,000 cfs in summer. The peaks exceeded 24,000 cfs 10 percent of the time and were below 5,000 cfs 10 percent of the time.

Tributary Inflows Below the Dam

Flows through the Grand Canyon are normally dominated by the water released from Glen Canyon Dam; tributaries below the dam such as the Paria River, Little Colorado River, and Kanab Creek (Figure 1.1) contribute less than

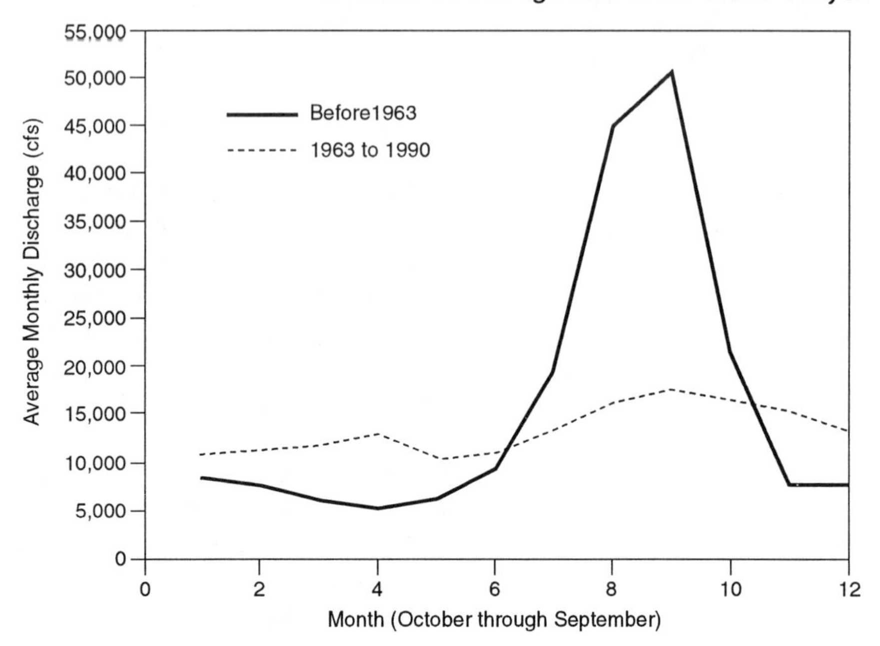

FIGURE 4.1 Comparison of monthly average discharge at Lee's Ferry before and after closure of Glen Canyon Dam in 1963 SOURCE: Bureau of Reclamation (1994).

2 percent of the average flow, as shown in Table 4.1. Small tributaries such as Havasu Creek and Bright Angel Creek contribute an insignificant amount of water. In a particular tributary drainage, however, thunderstorms can for a short period produce a major increase in discharge and sediment transport. For example, the 120,000-cfs extreme event shown in Table 4.1 for the Little Colorado River is an order of magnitude higher than the average release from the dam. In January 1993 such an unusual event moved a large amount of sediment from the Little Colorado River into the main river.

Hydrological Regimes During the GCES Research Periods

During much of the data collection phase for Phase I of the Glen Canyon Environmental Studies (GCES), releases from Lake Powell were unusually high (1983 and 1984). During GCES Phase II, hydrological conditions included years of both normal and low runoff (each of which resulted in the prescribed annual releases of 8.23 maf). Thus, the entire GCES study interval

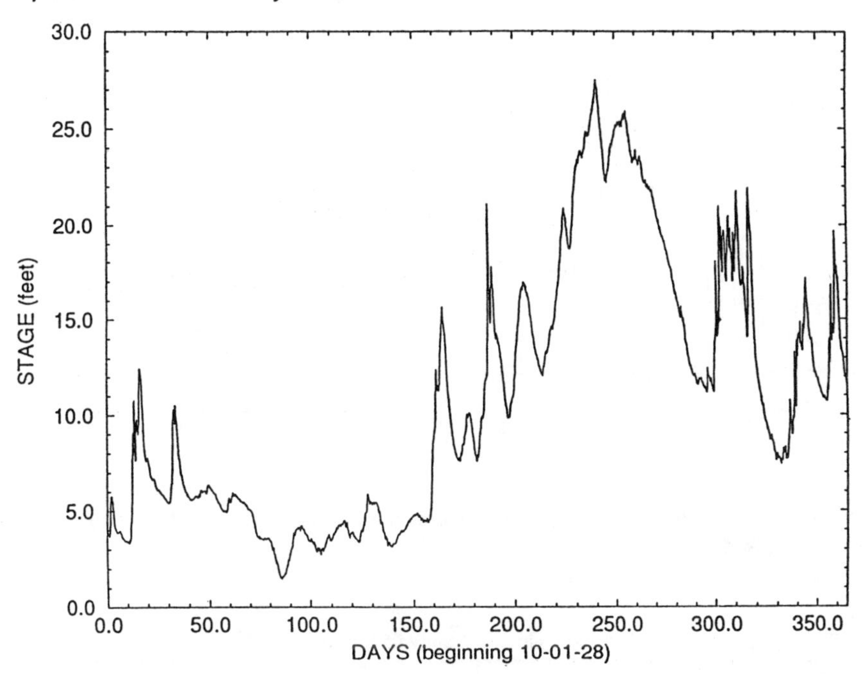

FIGURE 4.2 Example of fluctuations in stage height of the Colorado River prior to construction of Glen Canyon Dam (October 1928 through September 1929 at Colorado River near Grand Canyon gauge). SOURCE: D. Wegner, Bureau of Reclamation.

spanned a wide range of conditions for dam operation.

Prediction of future releases using data from the period of record since dam closure is difficult because much of the postdam historical record is biased by the large number of years during the filling period when releases were more restricted than they are now. During the initial filling of Lake Powell (1963 to 1980), water releases to the lower basin were reduced by 27 maf of reservoir storage plus 10 maf to 16 maf of bank storage. This was followed by an unusually wet period in 1983 and 1984 that resulted in flows approaching (and for a few weeks exceeding) turbine capacity for almost 2 years.

Travel Time Through Grand Canyon

During 1991, a dye study coinciding with experimental flows provided measurements of water velocities through the canyon at both steady and un-

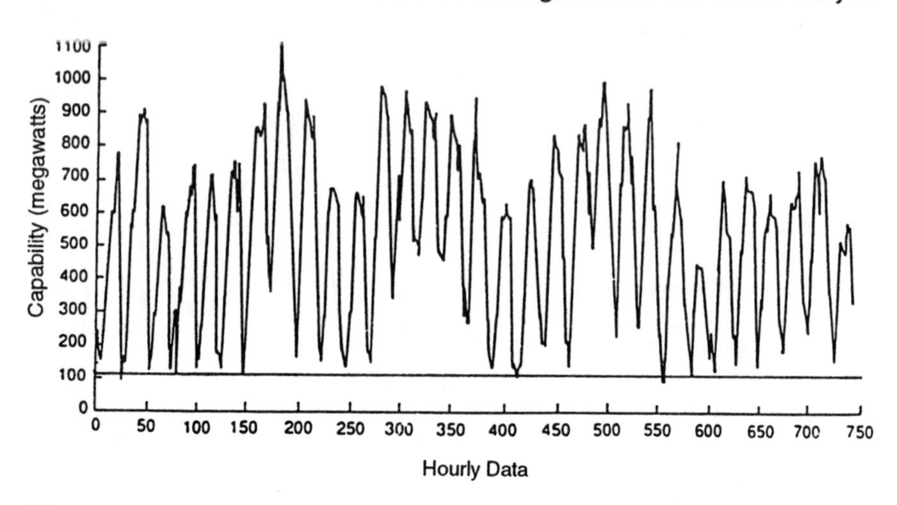

FIGURE 4.3 Example of Glen Canyon Dam operational fluctuations prior to interim flows, July 1982. SOURCE: WAPA (1988).

TABLE 4.1 Discharge of Tributaries in the Vicinity of Grand Canyon*

River	Annual Average (cfs)	Highest Annual Average (cfs)	Lowest Annual Average (cfs)	Extreme Event (cfs)
Colorado, Lee's Ferry	17,850	29,000	3,200	300,000
Paria River	29	65	11	16,100
Little Colorado	248	1,127	27	120,000
Kanab Creek	14	28	8	3,030

*From streamflow record 1922 to 1993.
SOURCE: USGS (1993).

steady flows (Graf, 1993). Wave velocities following rapid increases in flow rate were also documented and were used to verify calibration of an unsteady flow simulation model (Smith and Wiele, 1993). The flow rates shown in Tables 4.2 and 4.3 are from both the dye study and the flow simulation model. The average water flow rates were the same for steady flows (15,000 cfs) and unsteady flows with the same average rate (3,000 to 26,000 cfs).

As shown in Table 4.2, the wave caused by a rapid increase of water depth at the dam travels about two to two and three-tenths times the speed of the associated water mass. Wave celerity increases as a function of water

TABLE 4.2 Travel Time and Velocities Through Grand Canyon with Daily Flows Varying from 3,000 to 26,000 cfs and Averaging 15,000 cfs

		Water Travel Time		Wave Travel Time	
Location	River Miles	Days	mph	Days	mph
Lee's Ferry	0	0		0	
Little Colorado	62	1.4	2.2	0.6	4.3
Phantom Ranch	88	2.0	2.2	0.8	4.6
National Canyon	166	3.7	2.2	1.3	5.3
Diamond Creek	225	4.4	2.2	1.8	5.2

SOURCE: Graf (1993).

TABLE 4.3 Variations in Travel Time with Variations in Average Discharge (Lee's Ferry to Diamond Creek)

Average Discharge (cfs)	Time (days)	Velocity (mph)
5,000	10.0	1.0
15,000	4.4	2.2
30,000	2.9	3.4

SOURCE: Smith and Wiele (1993).

depth and change in depth. Water velocity increases with average discharge but less than linearly. For example, as shown in Table 4.3, as discharge increases by a multiple of 6 (from 5,000 to 30,000 cfs), mean velocity increases by a multiple of 3.4.

The travel times shown in Table 4.2 are for an average discharge (15,000 cfs) that is 36 percent greater than the annual average discharge (11,400 cfs). Thus, average travel times are somewhat longer than those shown in the table.

OPERATING RULES FOR GLEN CANYON DAM

Seasonal and Annual Operations

Formal Description of Operating Rules

The dam's daily operating rules are defined in terms of maximum and minimum releases and rates of change. Monthly releases are defined in a more complex way. The complexity derives from the fact that monthly releases involve conditional probability. They are recalculated monthly during the spring and early summer. The outcome of the calculations is a function of both snowpack and current storage level in Lake Powell. The monthly releases therefore differ every year and can be described only in statistical terms.

Monthly and Annual Releases

The annual release target for Glen Canyon Dam has always been 8.23 million acre-feet. This amount satisfies the Law of the River (Chapter 3) while maximizing the storage remaining in Lake Powell for use during future droughts. During wet years, when it appears from snowpack measurements that storage capacity will require more than this minimum release, the objective is to schedule monthly targets so that increased releases are distributed over several spring and summer months. This strategy reduces the likelihood of flood damage to the dam and, second, avoids bypass of hydropower turbines. Typical monthly release targets vary from about 0.5 to 1 maf. The smaller releases are in spring and fall; the largest releases are in summer and winter. Summer peaks accommodate demands for both hydropower and recreation, while winter peaks meet energy demands for heating.

The only change in monthly operating rules since GCES Phase I relates to the management of potential spills (bypass of turbines). During the unusually high runoff of 1983, a major spill resulted from the operating criteria, which then required that a minimum of 2.4 maf of storage be available on January 1. Releases were then estimated as necessary to fill the reservoir by July. The BOR's Colorado River Simulation Model (CRSM) estimated that this rule would result in a spill in 1 of every 4 years on average (BOR, 1986). Following the 1983-1984 floods, however, the criteria were revised. While the storage target on January 1 remains the same, the monthly release targets during years of high snowpack are now higher, particularly during spring months. Also, the estimates are made with a July storage target that is 0.5 maf lower than actual capacity. The new criteria have yet to be tested because the reservoir is now refilling after several years of drought, but the BOR's simulation model estimates that the frequency of spill with the new criteria will be 1 year in 20 on average.

The adjustments in monthly targets should cause no reduction in the probability that water can be delivered downstream as specified by the Law of the River. The annual target of 8.23 maf has the unusual characteristic of being both the probable minimum and the median annual release. It will be the minimum unless a long-term drought occurs that is much more serious than any in the 90 years of record. It is the median because during at least 50 percent of years the release has been, and will likely continue to be, no greater than 8.23 maf.

Daily and Hourly Operations

Daily Operations

Short-term operating rules prior to 1991 were designed almost entirely for maximizing the value of hydropower, except for a small accommodation to environmental resources in terms of minimum flow criteria. The releases were characterized by large daily fluctuations with peaks at turbine capacity of 31,500 cfs and minima of 1,000 cfs in winter and 3,000 cfs in summer.

In 1991 the daily operating rules were revised. These interim flow targets, which remained in effect as of September 1995, are:

- daily maximum releases not more than 20,000 cfs;
- minimum flows not lower than 5,000 cfs at night and 8,000 cfs during the day;
- change in release rate not to exceed 5,000, 6,000 or 8,000 cfs per day as monthly release targets vary from <0.6, 0.6 to 0.8, and >0.8 maf, respectively; and
- hourly changes in release rate not to exceed 2,500 cfs when increasing and 1,500 cfs when decreasing.

These interim flow rules are intended to reduce the adverse effects of dam operations on the environmental resources in Grand Canyon. The preferred alternative (modified low flow) that was selected in the operations EIS (BOR, 1994b) is very similar to the interim flow rules, but some modifications were made as a result of experience with operations during 1993 and 1994, as described below.

Experience with Interim Flows and Exception Criteria

The changes in daily operating rules have had no effect on the ability of the BOR to meet monthly or annual release targets because monthly release targets are totally independent of interim flow rules. It is, however, now much more difficult for the Western Area Power Administration (WAPA) to respond to hourly changes in energy load because of the constraints on both daily and hourly ramping rates (the rate of change in dam releases). Absolute enforcement of the revised operating rules would have decreased WAPA's ability to "meet system regulation needs, maintain transmission reliability, maintain operating reserve requirements, and serve firm load requirements" (WAPA, 1994). Therefore, in October 1991, WAPA and the BOR signed an interagency agreement. This exception criteria agreement allows WAPA to violate flow restriction rules not more than 3 percent of the time in any 30-day period (WAPA, 1994).

The range between the 20,000-cfs maximum and the 5,000-cfs minimum release limits appears to still allow substantial flexibility for response to changes in demand for hydropower. Because of the daily limits on ramping rates, however, the 15,000-cfs nominal range is reduced to 5,000 cfs in months with low-release targets and to 8,000 cfs in high-release months, as indicated in the following EIS preferred alternative section.

A recent example (one week in April 1995) of diurnal variations in both release rates and rates of change in release rates (ramping rates) are given in Figures 4.4 and 4.5. Clearly, the limiting parameter is ramping rate, not the 5,000- to 20,000-cfs bounds on release rate.

EIS Preferred Alternative

Operating criteria for the modified low-fluctuating-flow alternative, which was identified in the final EIS as the preferred alternative, are shown in Table 4.4.

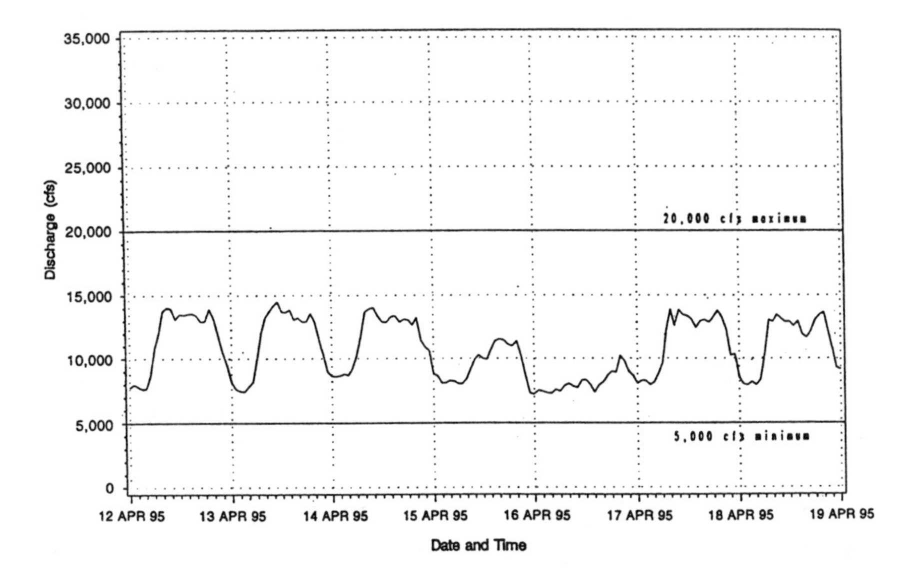

FIGURE 4.4 Hourly releases from Glen Canyon Dam for 1 week in April 1995. SOURCE: D. Wegner, Bureau of Reclamation.

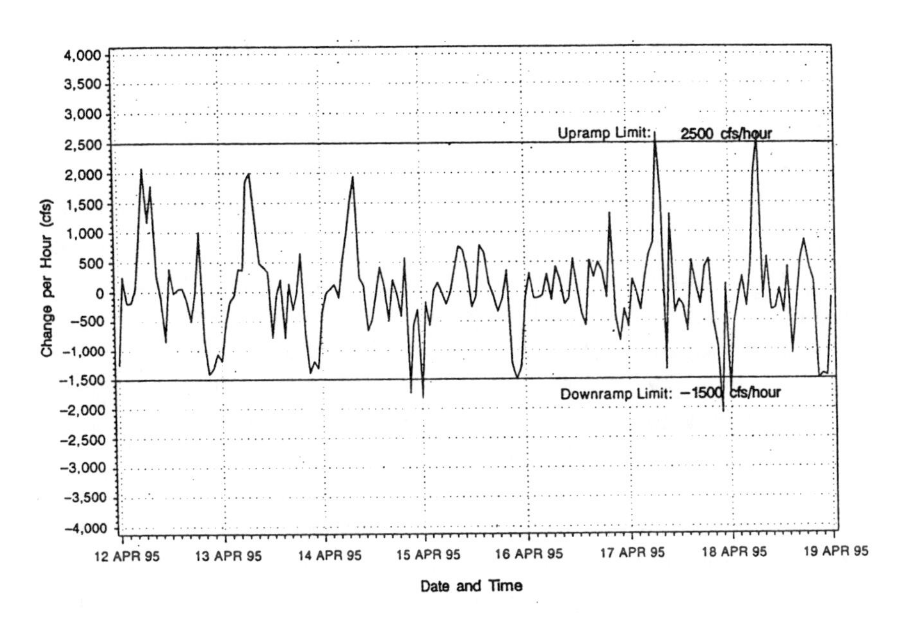

FIGURE 4.5 Ramping rates for release of water from Glen Canyon Dam for 1 week in April 1995. SOURCE: D. Wegner, Bureau of Reclamation.

TABLE 4.4 Operating Criteria for Modified Low Fluctuating Flow

Minimum Releases (cfs)	Maximum Releases (cfs)	Daily Fluctuations (cfs/24 h)	Ramp Rate[a] (cfs/h)
8,000/day	25,000/day	to 8,000[b]	4,000 up
5,000/night			1,500 down

[a]Rate of change in release from the reservoir.
[b]See Table 4.5.

The two significant changes from the interim flow criteria are the increase in maximum release rate from 20,000 to 25,000 cfs and the increase in ramping rate up from 2,500 to 4,000 cfs/h. The maximum flow rate occurs very infrequently, however, owing to the constraints on ramping rates. The down ramping rate (which is unchanged) was found to be much more important in terms of environmental impact than the up rate.

The range of variations in monthly release targets results in the approximate ranges in allowable daily fluctuation shown in Table 4.5, which are much less than the theoretically allowable variation between maximum and minimum daily rates. The preferred alternative also includes beach/ habitat-building flows as discussed below.

Beach/Habitat-Building Flows

Releases exceeding 25,000 cfs are included in the preferred alternative for most years (except when storage in Lake Powell is greater than 19 maf on January 1). These releases would occur during March at a steady rate of 33,200 cfs (power plant capacity) for 1 to 2 weeks. The purpose of these flows is to maintain physical habitat. Beach-building flows at rates higher than power plant capacity (45,000 to 52,000 cfs) are also part of the preferred alternative. They would occur with less frequency: 1 in 5 years, except when high runoff requires greater frequency.

The final EIS combines beach- and habitat-building flows. The combined flows are to occur either in May-June (high runoff) or in late summer during years when summer thunderstorms have added large amounts of sand from tributaries of the Colorado River below the dam. The flows are to be at least 35,000 cfs and are to last 1 to 2 weeks at a frequency of 1 in 5 flood years

TABLE 4.5 Fluctuation Range Experience Under Interim Flow Criteria

Monthly Release Volume (acre-feet)	Minimum-Flow day (cfs)	Minimum-Flow Night (cfs)	Allowable Daily Fluctuation (cfs)
<600,000	8,000	5,000	5,000
600,000 to 800,000	8,000	5,000	6,000
>800,000	8,000	5,000	8,000

SOURCE: Bureau of Reclamation EIS.

except when high runoff requires that they be more frequent. The actual size of these flows is to be determined from the results of an experimental flood exceeding 35,000 cfs, which has yet to occur.

Operational Constraints on Experimental Floods

Beach-building flows would exceed turbine capacity (33,200 cfs). A release rate of 45,000 cfs can be achieved by using the river outlets, which have a capacity of about 15,000 cfs (actual capacity depends on reservoir storage level). Releases above 50,000 cfs would also require some flow through the spillway tunnels, which are controlled by radial gates (50 feet high). This can be done only during years when the lake level is above the bottom of the radial gates (elevation 3,648 feet above sea level). This elevation corresponds to 17 maf of storage, which the lake level is expected to exceed except following extended droughts. BOR considers use of the spillway undesirable on a routine basis, because it is considered to have the shortest useful life of any of the components of the dam, even after being modified to prevent cavitation following the 1983 flood.

WATER SUPPLY ABOVE THE DAM

Water Balance in Lake Powell

All predictions of future probabilities for release or spill of water from Glen Canyon Dam are based on assumptions that are inherent in the mass balance equations of the CRSM. This model uses gauge records of inflow to Lake

Powell, releases through the turbines, and gauge data for the Colorado River at Lees Ferry. Computation of mass balance also requires estimates of evaporation and change in bank storage (these variables are not measured). The mass balance equation used is

$$S_{t+1} = S_t + Q^i - Q^0 - E - .08(\Delta S),$$

where S_{t+1} is ending storage, S_t is beginning storage, Q^i is flow into the reservoir, Q^0 is flow released, E is evaporation, and ΔS is change in bank storage.

The CRSM estimate of evaporation appears to be much lower than predicted by various researchers (Dawdy, 1991; Hughes, 1974). Also, change in bank storage is estimated as a constant 8 percent of the annual change in storage, regardless of reservoir level, even though bank storage would be expected to vary in relation to reservoir level. The recent drought (1987-1992) lowered the reservoir to an unprecedented extent. Data on water balance over this interval should provide a way to improve the bank storage estimate. Mass balance estimation could be improved through an analysis of the data on flow and reservoir stage, including the recent drought period, combined with a corrected estimate of evaporation over the same period.

There is also an apparent discrepancy between reservoir release rate as measured by the energy generated at Glen Canyon Dam and the flow rate measured at Lee's Ferry. Part of this difference is due to seepage from the reservoir, but an analysis of possible errors in both approaches should be made so that seepage can be determined more accurately. Because of the recent large drawdown, it should now be possible to improve estimates of seepage as a function of storage in the reservoir.

Upper-Basin Depletions

During the 1980s, projections of future release frequencies by the CRSM model assumed that consumptive use in the upper basin would progressively increase, which would reduce releases to the lower basin over the next 60 years (Figure 4.6). A large number of projects for the upper basin have been authorized by Congress, and the CRSM model has in the past incorporated the assumption that these states will eventually use their entire share of the river (6.0 maf). It is now clear, however, that most of these projects will not

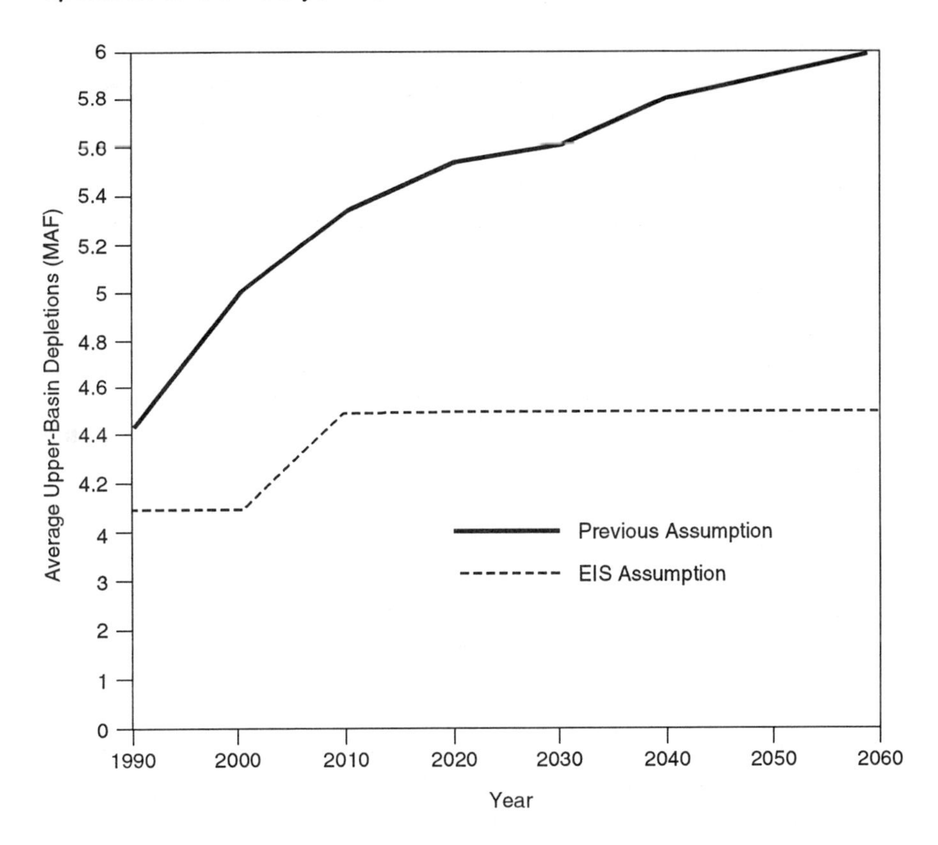

FIGURE 4.6 Comparison of previous and current projected depletions from Colorado River above Glen Canyon Dam. SOURCE: Bureau of Reclamation EIS.

be built because of their adverse environmental effects and new criteria for allocating the costs of irrigation projects. Therefore, the CRSM estimate of upper basin depletions used for the EIS model runs was set at 4.5 maf per year on average over the next 50 years (Figure 4.6). This makes a significant difference (a decrease) in the number of years during which the annual release from the dam will be at the minimum level.

Evaporation Loss

Since the completion of Glen Canyon Dam, the Colorado River has reached the sea only during 2 years of unusually high runoff years (1983-

1984). This implies that, on average, the 12 maf of water produced by the watershed leaves the system in one of three principal ways: (1) as evapotranspiration from irrigated lands (by far the largest quantity), (2) as effluent from urban sewer treatment plants (mostly in Southern California), or (3) as evaporation in transit (mainly from reservoirs).

The BOR operating plan for 1994 estimated the total evaporation loss from the entire storage system as 1.6 maf, or 13 percent of the total resource. For Lake Powell the estimate is 507,000 acre-feet (BOR, 1994a). This is from an average summer-season surface area of 139,000 acres with an estimated annual evaporation of 3.65 feet. This may well be an accurate estimate of evaporation loss in calculating upper-basin depletions for the purpose of evaluating water rights. It is not, however, the total evaporation from the reservoir. It is calculated as the difference between the net evaporation (corrected for precipitation) from the open water surface of Lake Powell and the evapotranspiration from the land surface that was inundated by the reservoir (mostly from phreatophytes). This net loss is calculated as the difference between stream flow gauges (corrected for minor ungauged streams) and the estimated flow at Lee's Ferry without the dam, as shown by gauge data at Lee's Ferry prior to 1963. This continues to be projected as the loss from Lake Powell. While this may provide a correct estimate of depletion caused by dam construction, it is much less than the evaporation that would be used in mass balance calculations for the river simulation model. Dawdy (1991) and Hughes (1974) have estimated the correct net evaporation (corrected for rainfall) as about 5.3 feet for Lake Powell.

The significance of the conceptual error in estimating of evaporation can be demonstrated as follows. Since 1964 when the lake began filling, the average storage has been about 15 maf. The surface area at this elevation is 115,000 acres, which suggests an average annual net evaporation during the past 30 years of 609,500 acre-feet at a rate of 5.3 feet per year. The BOR estimate at this same lake level is 419,750 acre-feet. This is a difference of 189,750 per year, or 5.7 maf in 30 years.

The BOR staff recently calculated all of the monthly corrections to the mass balance equation used by the simulation model during the past 30 years. The total was 6.5 maf of loss unaccounted for by the model. A correct evaporation estimate would account for 88 percent of this error. BOR, however, chose to make the correction as bank storage and added a one-time correction of 6.5 maf to the bank storage estimate in October 1993.

It is correct to show the open-water minus the predam evapotranspiration as the "loss" charged to Glen Canyon Dam for upper-basin water depletion calculations. It is incorrect, however, to continue to use this same quantity in the reservoir operating plan mass balance calculations. This introduces an annual error that is greater than the entire proposed Central Utah Project diversion from the river.

Storage in Lake Powell

There is about 2 maf of dead storage in Lake Powell. Figure 4.7 displays active (not total) storage above the elevation of the river outlet pipe centerlines (elevation 3,374). The curve in the figure displays the probability that the lake level will be above any given storage volume or elevation (BOR EIS draft, Appendix B, p. B-154). For example, the figure indicates that 70 percent of the time water is expected to be above the bottom of the radial gates—which is the minimum elevation for controlled releases exceeding the turbine capacity (as necessary for an experimental flood or 50,000 cfs).

The elevation of the eight turbine intake pipe centerlines is 3,470 feet above sea level. The elevation below which the turbine operation would cease is 3,490 feet above sea level. This 20 feet of head above the turbine intakes is necessary to prevent air from being sucked into the turbines. Below an elevation of 3,490 feet, the only way to release water would be from the river outlets, which are at 3,374 feet. Figure 4.7 suggests the probability is zero that water level will fall below the elevation at which the turbines would have to be stopped. This frequency analysis is based upon the Bureau's simulation model CRSS, which as previously stated, underestimates evaporation from the lake. If corrections for the more accurate evaporation losses were made, the exceedance line in Figure 4.7 would be lowered (except at the left end).

Flood Control

Current Rules Related to Floods

Currently, 2.4 maf of storage space is reserved for floods on January 1 of each year. This flood storage space is gradually reduced to 500,000 acre-feet in June, when the runoff peak has begun to decline. In addition, the expected quantity of peak season runoff is increased by a safety factor that is highest in January (4.98 maf) and declines as uncertainty declines to 2.13 maf on June

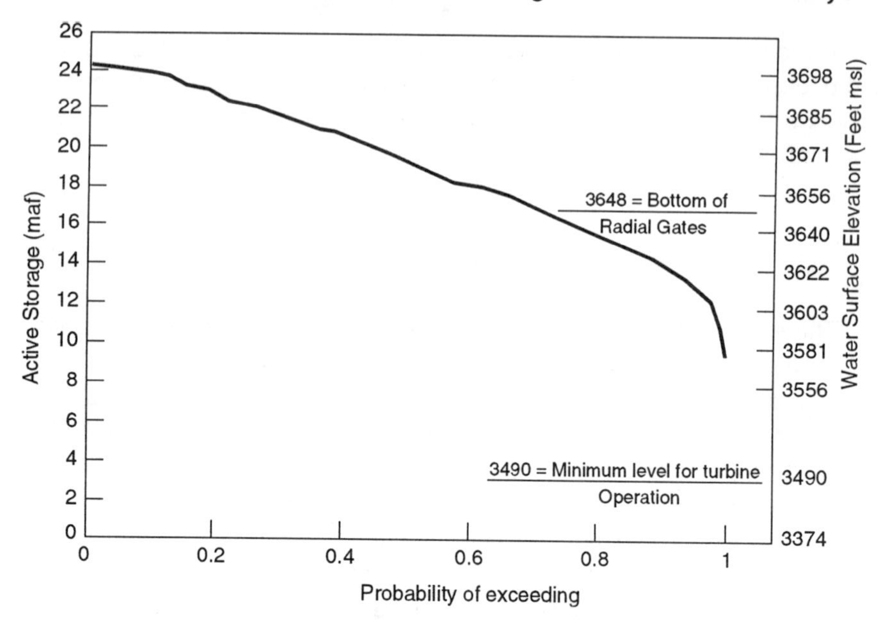

FIGURE 4.7 Frequency analysis of reservoir water levels at beginning of water year (October 1), as predicted by the CRSM.

1. These practices are estimated to prevent floods exceeding turbine capacity of 33,000 cfs in 19 of 20 years on the average.

Flood Control Changes Recommended in EIS

Additional flood control measures have been recommended in the operations EIS. These measures would reduce the frequency of floods exceeding 45,000 cfs to 1 in 100 years. There are two possible ways to achieve this reduction in flood frequency, as described in the EIS: (1) raise the top of the spillway radial gates by 4.5 feet, which would provide 0.75 maf of additional flood storage, and (2) change the releases to target a maximum reservoir content of 23.3 maf during spring months until the runoff peak has passed. This is 1 maf less than the previous target level.

Reregulating Dam

The conventional approach to mitigating the environmental effects of daily fluctuations below hydropower dams is to construct a small dam below the main dam. The purpose is to equalize, or at least reduce, the daily variations in releases from the dam. The amount of storage behind a secondary dam need only be a fraction of the volume of daily releases. In the case of Glen Canyon Dam, a reregulation dam could be located about 17 miles below the main dam, or 1/2 mile above Lee's Ferry in the Glen Canyon recreation area, but outside the national park. Such a dam would raise the water level by about 20 feet at the reregulation dam (BOR, 1994b, p. 49). This would convert the reach between the two dams from a river (with a prime trout fishery) into an impoundment within which fishing would likely not be allowed because of the rapid changes in stage. The principal benefit of such an arrangement is to allow river releases at approximately the optimum level for environmental purposes while allowing the main dam to follow electrical demand and thus generate maximum hydropower revenues.

The economics of the reregulation concept are interesting in that the reduction in the annual value of hydropower estimated by the EIS for the preferred alternative ($30 million annually, estimate from Bureau of Reclamation) would pay the capital cost of constructing the reregulating dam in a very few years. A thorough analysis should be done if a reregulation dam below Glen Canyon Dam is to be considered. A reregulation dam below Glen Canyon Dam would allow both maximization of hydropower value and optimal releases for environmental objectives—thereby reducing many of the concerns centered around operation of the dam. The disadvantage of building a reregulation dam is that it would inundate another portion of the canyon.

Summary

The changes in operating rules for Glen Canyon Dam resulting from the Grand Canyon Protection Act and the EIS preferred alternative have no effect on the dam's long-term operation (monthly and yearly releases). These changes affect only the way in which daily average releases are distributed hourly.

While the range between revised maximum and minimum allowable release rates is still quite large (5,000 to 25,000 cfs), the possible daily fluctuations resulting from the preferred alternative rules are much less—5,000 to 8,000 cfs depending on the monthly release volume target. This smaller range is due to ramping rate limitations that now dominate the allowable short-term variations in dam releases.

The predam flows through the Grand Canyon frequently experienced very substantial daily fluctuations due to local storms in the tributaries—a flow regime quite different than constant daily flows.

The BOR's Colorado River simulation model has a conceptual error in the way in which evaporation losses from Glen Canyon Dam are calculated. As a result, the losses are significantly understated. The method of simulating both evaporation volumes and bank storage volumes should be improved.

A reregulating dam below Glen Canyon Dam may allow both maximization of hydropower value and optimal releases for environmental objectives—thereby reducing many of the concerns that have driven the controversy over the dam's release patterns.

RECOMMENDATIONS

1. An analysis of data (including the recent drought period) on reservoir inflow, outflow, evaporation, and reservoir stage is needed to improve the estimate of bank storage in Lake Powell.
2. An analysis of the comparison between releases from the dam (based on energy generated) and the flow measured at Lee's Ferry should be made so that seepage can be determined more accurately.
3. For its mass balance simulation model, the BOR should use actual open-water evaporation rather than depletion (the difference between the open-water evaporation and the predam evapotranspiration).
4. If the option to mitigate environmental effects of daily fluctuations below Glen Canyon Dam includes building a reregulation dam, a thorough analysis of its costs and possible environmental impacts should first be completed.

REFERENCES

Bureau of Reclamation. 1986. Special Report, Colorado River Alternative Operating Strategies for Distributing Surplus Water and Avoiding Spills. BOR, Denver, Colo.

Bureau of Reclamation. 1988. Glen Canyon Environmental Studies. Final Report, U.S. Department of the Interior, Washington, D.C.

Bureau of Reclamation. 1994a. Annual Operating Plan for Colorado River Reservoirs, 1994. U.S. Department of the Interior, Washington, D.C.

Bureau of Reclamation. 1994b. Operation of Glen Canyon Dam. Environmental Impact Statement, U.S. Department of the Interior, Washington, D.C.

Dawdy, D.R. 1991. Hydrology of Glen Canyon and the Grand Canyon. In Colorado River Ecology and Dam Management. Washington, D.C.: National Academy Press.

Graf, J. 1993. GCES Travel Time and Dispersion. Draft Report. Tucson: U.S. Geological Survey.

Hughes, T. C. 1974. Water Salvage Potentials in Utah, Vol. 1. Utah Water Research Laboratory, PRWA22-1. Logan: Utah State University.

Smith, J., and S. Wiele. 1993. GCES Sediment Transport Study Draft Report. Boulder: U.S. Geological Survey Water Resources Division.

U.S. Geological Survey. 1993. Water Supply Records, Arizona.

Western Area Power Administration. 1994. Salt Lake Area Integrated Projects Firm Power Proposed Rate Adjustment Brochure, Salt Lake City.

Western Area Power Administration. 1988. Salt Lake City Area, Analysis of Alternative Release Rates at Glen Canyon Dam.

5

Sediment and Geomorphology

WHY SEDIMENT AND GEOMORPHOLOGY ARE IMPORTANT

Sediment is an integral part of the Grand Canyon river system. Sand entering the Colorado River from the major tributaries moves through the canyon with the river and is deposited as bars. When the bars are above the water surface, they become the beaches on which the river runners camp. Occasional movements of coarse substrate (debris flows) from tributaries create the rapids that the boaters thrill to travel through. Sand and debris flows provide habitat for riparian and aquatic communities. Thus, an understanding of the processes that create bars, beaches, and rapids is a necessary part of the Glen Canyon Environmental Studies (GCES).

Amount of Sediment Is Not a Problem in the Grand Canyon

There is adequate sediment, of which sand is the dominant constituent, entering the Colorado River below Glen Canyon Dam to maintain bars and beaches. In fact, sand would not accumulate in the canyon if flow from the dam were constant (BOR, 1993, p. 90) but would collect in the channel and in low bars, with an elevation just below the elevation of the water surface. Thus, the problem of sand in the Colorado River is not the amount of sand but rather its distribution.

Management of Sediment

Maintenance of beaches above mean water level requires some procedure for raising the water level temporarily for beach-building flows. The amount of discharge (height of water in the channel) will determine the height of the beaches that are built by beach-building flows. Furthermore, beach building will be most successful when there is a large supply of sand near the upper end of the Grand Canyon. A sand budget must be maintained that shows the amount and distribution of sand.

As we will discuss later in this chapter, the critical reach for sand management is between the Paria and the Little Colorado rivers. Research and data collection should concentrate on that reach of river. Of secondary concern is the reach from the Little Colorado River to Phantom Ranch. Much less important in terms of sand management is the reach below Bright Angel Creek.

Sand cannot be lifted onto beaches without occasional high flows (Andrews, 1991). The amount, duration, and scheduling of high flow are elements of management decisions that must be based on good scientific data plus reliable physically based models. In general, higher and longer flows create higher beaches. Beach-building flows should coincide with the presence of high amounts of sand in the reach from the Paria to the Little Colorado rivers; this sand will be lifted to beaches downstream by the high flows.

At present, the sand supply from the Paria River is less well known than that of the Little Colorado River. The Paria should be closely monitored so as to reduce the uncertainty in the river's sediment transport record. A study of the relationship of accuracy to frequency of sampling should be undertaken for the Paria.

Rapids

Rapids in the Grand Canyon are created by debris flows from minor tributaries throughout the canyon. Sands and gravels are eroded from the debris flows, leaving behind the cobbles and boulders. The debris flows narrow the area of flow and thus create the rapids.

The size of the materials removed from the debris flow deposits depends on the peak flows that pass around and over the debris. Before the closing of Glen Canyon Dam, the rapids were reworked by large annual floods and by occasional massive floods. Thus, the rapids might constrict the channel, but the amount of constriction was periodically reduced by annual and occasional major floods.

Without inadvertent spills, as occurred in 1983, or deliberate flood flows, Glen Canyon Dam would cause the rapids to increase in size. Without flood flows, there would be no mechanism to remove the large material in the debris flows (Cooley et al., 1977; Dolan, 1981; Kieffer, 1987). Each time a debris flow occurs at a site of an existing rapid, the new debris will further constrict the channel. Over a period of time, rapids could become impassable, and river runners would need to portage around them. Thus an understanding and monitoring of the debris flows and their removal mechanism were integral parts of GCES.

Sand as Substrate

Sand deposits are a substrate that supports riparian and aquatic organisms in the canyon river corridor. Backwaters behind the beaches provide habitat for warmwater-adapted fishes, such as the humpback chub (Chapter 6). Releases from Glen Canyon Dam are cold and swift, and the backwaters reduce current velocity and may also give the water a chance to warm. The bars and beaches support vascular plants and thus provide habitat for terrestrial animals. Thus, an understanding of the physics of the sand system and of the geomorphology of the canyon are important for understanding the occurrence and abundance of the flora and fauna throughout the canyon.

In the past, sand has covered archeological sites and thus has protected them. Periodic erosion of the sand through local flooding was offset by deposition of new sand from the annual floods of the Colorado River. Without the high annual floods and occasional rare large floods in the Colorado River, local erosion on the beaches and bars will not be offset by deposition without some inclusion of high flows in the management plan for the dam. Local runoff now creates gullies and thus exposes ancient sites. Only flood flows can reverse this trend (Hereford et al., 1991).

SEDIMENT STUDIES OF GCES PHASE I

GCES began in 1982 in response to concerns about the effects of Glen

Canyon Dam on the resources of the Grand Canyon. The planners of Phase I of GCES recognized that sediment is a major resource of the Grand Canyon river corridor. The daily fluctuations of flow from Glen Canyon Dam (Chapter 4) were thought to erode the beaches and thus diminish the resources for river runners as well as the native fishes and riparian biota of the canyon.

Almost as soon as GCES Phase I was organized, exceptionally high runoff occurred (in 1983). The resulting unusual spill from the dam radically altered the canyon and its beaches in ways that were difficult to anticipate. Thus, the high flows disrupted the original plans for GCES Phase I.

Bureau of Reclamation Model

GCES Phase I included studies of sediment movement through the Grand Canyon and emphasized modeling of flow and sediment transport. Modeling was undertaken by the Bureau of Reclamation (BOR), which planned to adapt a preexisting steady-state program to model the flow through the canyon (Randle and Pemberton, 1987). The movement of flood waves was modeled by a series of steps involving different discharges. Thus, the flow was allowed to vary, but the shape of the flood wave was held constant.

The BOR modeled sand movement by using the hydraulics developed by the water-routing model with calibration to sand discharge relationship at various control points in the canyon. The gauge below the Little Colorado River and the long-term gauge at Phantom Ranch (Bright Angel Creek) were to be the anchors of the model, and other temporary gauges were used to supplement the data base.

There were two major flaws in the BOR approach to modeling in GCES Phase I. First, a steady-state model cannot predict the attenuation of a flood wave as it moves down a river. The second major flaw was that the sand discharge relationships at several measurement sites in the canyon could not be differentiated from each other. Thus, the model was necessarily calibrated to identical relationships at different sites, which led to the conclusion that the sand input and output would be the same throughout the reach of the canyon modeled. Thus, there was no basis for estimating the variation in storage of sand within reaches of the canyon.

GCES Phase I monitored the beaches, their geometry, and their volume. There was, however, no strategy for relating these results to the water- and sediment-routing model.

Early Assumptions Concerning Sediment Movement

GCES started with the assumption that beaches were being eroded throughout the Grand Canyon and that the cause was fluctuating flows released from Glen Canyon Dam. The judgments stated above were those of the river boatmen, but they strongly influenced the GCES researchers. These assumptions influenced the decisions concerning what should be studied, how it should be studied, and, more importantly, the conclusions from GCES Phase I. Part of the learning experience of GCES Phase I was to set research goals as scientific hypotheses to be tested rather than assumptions to be verified.

Coordination and Communication

"Unfortunately, the critical linkage of sediment/water flow in the main channel was pursued predominantly as an exercise in its own right, largely divorced of concerns about sediment sources and sinks and with inadequate attention to modeled sediment movement to beaches, riparian habitats, and so on" (NRC, 1987, pp. 88-89)—so stated the National Research Council (NRC) committee in its assessment of the sediment work in GCES I. Good work was performed and excellent data were collected, but there was little coordination among the different elements of the research team. Elements were added to the plan as time showed that the original plan would not provide sufficient information. The spill of 1983 radically altered the work plan but did not lead to full integration of the study team. New projects were added to study newly perceived problems, but each project remained essentially an independent entity. There was little coordination of results and little exchange of information among research teams. The water and sand modeling was not related to the beach studies, and there was no mechanism for the results of one to be integrated into the work plan of the other.

Leadership for GCES

GCES I was organized around the capabilities and approaches of governmental agencies. Each agency pursued its work independently, with little sense of overview. This resulted partly from the fact that each agency was considered an expert in its field, and the GCES leadership was not sufficiently empowered to overrule the judgment of individual agencies for the benefit of GCES (Chapter 2).

For sediment research the agencies were asked to propose projects that would address the general goals of GCES. There was no a priori statement of tasks, with work plans, and with requests for agencies to perform the work as designed by GCES. Thus, for example, the BOR worked on its water and sediment model without anyone asking how the model and its results would fit into the solution of the problems to be addressed by GCES I.

NRC Committee Recommendations After GCES I

At the end of GCES I, the NRC committee made several suggestions concerning future GCES work. Several suggestions were particular to sediment work. Others, though more general, had direct bearing on it. The particular sediment-related suggestions were summarized as follows: "Future work by the Department of the Interior should seek to look for connections between research disciplines in the planning phases of the study, initiate studies of tributary processes because they are the main source of sediment in the Colorado River main stem, include in future hydrologic research empirical approaches as well as modeling approaches, link sediment studies to biological and hydrological monitoring and research, and institute geomorphic studies to supplement the hydraulic studies of the Colorado River system in the Grand Canyon" (NRC, 1987, p. 8). The GCES team implemented some of these suggestions. In particular, the tributary sources of sediment were given more emphasis. Both empirical and physically based modeling replaced the original modeling effort. The general problems of coordination and integration persisted. Different groups collecting similar data at a site did not always communicate their findings or plans to each other. There seemed to be no understanding that the data collection could be better coordinated, so that the overall results of GCES could be achieved with less effort. There seemed to be little attempt by researchers to query each other before a river trip to determine whether there were data that could be collected for general use or that could be collected in a slightly different manner or format for the benefit of others. As an example, the slopes of the bar faces were studied, but bar topography was documented separately and without acknowledgment of the values of computing bar face slopes for use by others.

WHAT IS NEEDED CONCERNING SEDIMENT AND GEOMORPHOLOGY

Sediment Budget

The perception of the problems concerning sand in the Grand Canyon changed over the life of GCES. One of the concepts that instigated GCES was the belief that fluctuating flows from the Glen Canyon Dam were causing a loss of the sand bars, which are the camping beaches on which the public depends in their trips through the canyon. It was thought that steady flows would reduce the removal of sands and improve the condition of the canyon as a whole.

About 790,000 tons of sand enter the Colorado River annually through the Paria, and another 1,600,000 tons enter through the Little Colorado River. Kanab Creek contributes 300,000 tons per year, and other minor tributaries and debris flows from side canyons contribute about 700,000 tons of sand per year. Thus, there is on average, over 3 million tons of sand delivered per year to the river corridor (BOR, 1993, Appendix D). If the releases from Glen Canyon Dam were constant, about 11,400 cubic feet per second (cfs) would correspond to a release of 8.25 million acre-feet (maf) per year as specified by the Law of the River, and 15,200 cfs would carry the average inflow to Lake Powell (11 maf per year). A steady flow of 15,200 cfs would carry about 550,000 tons of sand per year to the Little Colorado River, and some 1.13 million tons per year past Phantom Ranch and out of the canyon into Lake Mead (BOR, 1993, Appendix D). Thus, with steady flows, sand would accumulate in the canyon (almost 2 million tons per year). It would accumulate in the main channel, however, and not on the beaches and bars.

Wave action eventually would cause all bars to erode to the elevation of the water if flows were steady. If steady flow persisted, more of the bottom of the channel would be covered with sand and a new equilibrium condition would develop that would carry the sand through the Grand Canyon with bars at water level. The only variation in this condition would be temporary, following storm flows from the Paria and Little Colorado rivers, which might deposit sand at higher elevations.

Sediment Budget by Reach

The problem of sand management varies in different parts of the canyon.

In the first 16 miles, from Glen Canyon Dam to Lees Ferry and the mouth of the Paria River, there is almost no sand inflow (38,600 tons per year from minor tributaries). Therefore, the sand storage in that reach is declining. Sand transport past the mouth of the Paria therefore is a net loss to the reach above. This can be ameliorated only by sand augmentation, which presently seems unlikely (Chapter 2). Otherwise, all sand in the first 16 miles eventually will move downstream, and there will be no beaches in that reach.

In the next 61 miles, between the mouth of the Paria and the mouth of the Little Colorado, the only significant sand supply is from the Paria. The distribution of sand in that reach is dynamic, and the capacity of the water much exceeds that required to move the sand downstream. This is the critical reach, and the problem of sand management is to keep as much sand as possible on the beaches and out of the main channel.

Below the Little Colorado River, almost the total supply of sand is available in the Colorado River. The first 26 miles, between the Little Colorado to Bright Angel Creek, will, however, have periods of beach erosion when sand supply to the Colorado River from the Little Colorado is below normal. Beach erosion should be considered normal during periods of sand supply deficit, and sand accumulation should be expected during periods of high sand transport from the Little Colorado.

Any strategy that improves the reach from the Paria to the Little Colorado probably also will improve the reach just downstream from the Little Colorado. This is because the flows from the Paria and the Little Colorado are correlated; high flows and high transport of sand usually occur from both in the same year. Thus, when there is a supply in the upper reach to be managed, there usually is a supply in the reach just below the Little Colorado also.

The Paria, which is much smaller than the Little Colorado (Chapter 4), may not increase the main river flow sufficiently to transport the sand that it delivers. Thus, much of the sand contributed by the Paria will tend to remain in the channel immediately below the river's mouth. In contrast, the peak flows from the Little Colorado combined with flows from the Paria and with dam releases often exceed the power plant capacity of 31,500 cfs from Glen Canyon Dam and will deposit materials on the beaches throughout the system downstream from the Little Colorado. In particular, if the Little Colorado peak coincides with high flow from Glen Canyon Dam, a sizable peak of discharge and transport can occur.

The most critical piece of information in the sand budget will be the amount of sand in the main channel between the Paria and the Little Colorado. There should be some decision criterion that triggers the release of beach-building flows sufficient to deposit sand from the main channel onto the beaches in the reach above the Little Colorado. The amount of sand in storage in that reach should be a part of the triggering, as well as the water stored in Lake Powell. When both are above normal, beach-building flows can and should be planned.

Sand Distribution Between Main Channel and Beaches

Beaches form where eddies can remove sand from the main channel. Beaches often form below rapids, which create eddies downstream. Knowledge of the mechanism of eddy formation, the variation of the mechanism with discharge, and the process for transfer of suspended sand across the eddy fence (a vertical plane that divides an eddy field from other parts of the channel flow that are moving in a downstream direction) into the eddy field (an area of flowing water where motion occurs in a circular fashion or in a reverse direction to the rest of the channel) is needed.

Preliminary evidence seems to indicate that the transfer of suspended sand from the main channel across the eddy fence is fairly rapid (Dawdy, personal communication, 1993). Therefore, the bars will not be eliminated as long as there is suspended sand in the main channel. Eventually, the bar within an eddy will develop to a height about equal to the steady water level. Thus, a strategy must be developed not only for predicting the movement of sand across the eddy fence but also for predicting the deposition of sand on the beaches at higher elevations during beach-building flows.

The river stage heights of the beach-building flows will determine the height of the camping beaches. The greater the flows, the higher the beaches will be and the greater the amount of sand that will be stored on the beaches. An understanding of the eddy system and its role in beach building is an integral part of managing the resources of the Grand Canyon.

Flood Waves

Peaks attenuate as they travel downstream. The attenuation can be predicted quite well with the models developed by the USGS for GCES (Smith and Wiele, in press). The flood wave changes shape as it moves downstream, and the travel time of the wave varies as it moves downstream (Figure

5.1). The leading edge of the wave steepens as the wave moves downstream, whereas the falling edge becomes less steep. As peaks attenuate and peak stages are lower, the beaches that the wave can build will be at a lower elevation. Also, brief peaks deposit less sand. An understanding of this interaction between sand deposition and beach-building flows should be developed, so that the sand resource can be managed throughout the system.

U.S. Geological Survey Effort on Sediment Modeling

The U.S. Geological Survey (USGS) has developed a water-routing model that can predict the transformation of the flood wave down the channel. The steepening of the wave front, flattening of the back of the wave, lowering of the peak, and increase in discharge at the trough can be modeled. Thus, for any time at a given site, the USGS model can be used to predict the elevation of water level, given a history of discharge (Smith and Wiele, in press).

The linking of discharge to sand transport is more tenuous. The USGS uses a reach-averaged cross section to model the stage. It uses the percentage of bottom covered by sand as input to its sediment transport prediction. There seems to be no periodic updating of the cross section as sand is predicted to accumulate or erode. The model is calibrated to the sediment collection sites in the canyon, which are the source of information on percentage of the bottom covered by sand in the reaches upstream from each measurement site. The model is an improvement over the steady-state model of GCES I, but it still is the weak link in the modeling needs for managing sand.

USGS Effort on Eddy Modeling

Eddy modeling is farther along than main channel modeling, although it was less well understood when GCES began. There has been more emphasis on this aspect of sediment modeling than on any other, and there have been more workers attacking the problem from different aspects. Also, there seems to be good communication among workers studying geomorphic and mathematical aspects of eddy formation, circulation, and bar building.

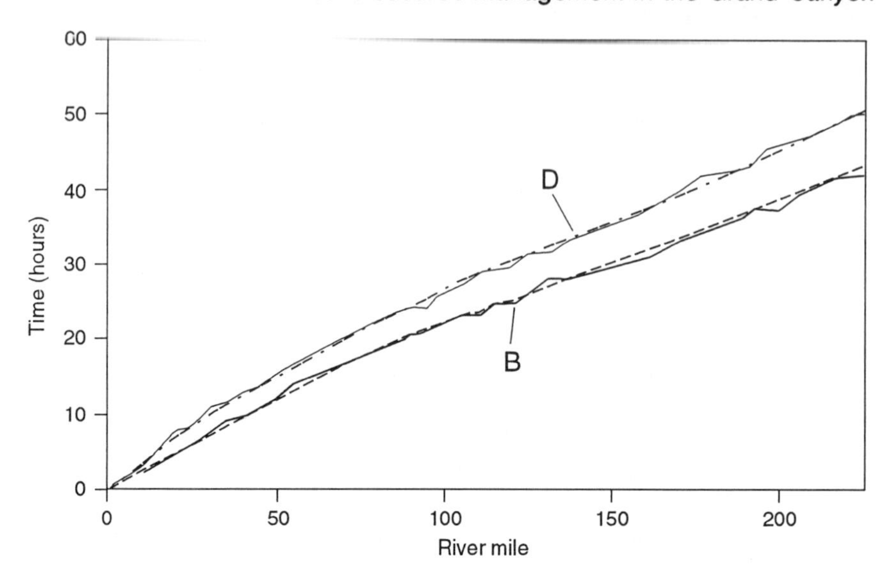

FIGURE 5.1 Travel time as a function of river mile for discharge waves during Research Flows B and D (solid line) compared to predicted travel times [dashed line with single dots (b) and dashed line (d)]. The smaller-amplitude discharge wave of Research Flow B had a lower phase speed and therefore is the lower of the two solid lines. SOURCE: U.S. Geological Survey (1991).

Linking of Sediment Modeling and Eddy Modeling

At the present time, there still is a need for linking the main channel water and sand discharge with the eddy system and the growth of bars. The deficiency at this time appears to be more in the main channel sand modeling than in the eddy modeling. Some first-approximation results based on the connection of the sand transport model and the eddy circulation model should be developed to predict the growth of bars.

NAU Beach and Bar Studies

Northern Arizona University (NAU) has an extensive program on mapping of bars (Beus et al., 1994). There seems to be little attempt at quantitatively relating these measurements to the building of bars or to using the data for scheduling beach-building flows. Various data collection efforts should have been coordinated with each other. The types of data collected should have been determined by GCES management, so that the NAU data could have been used by other researchers.

WHERE WE STAND NOW

The GCES has added considerably to our understanding of the physics of sand movement in the Colorado River below Glen Canyon Dam, but there is still much that needs to be understood. Among the GCES findings related to sediment are:

- There is a sufficient amount of sand reaching the Colorado River to maintain the beaches. The problem is one of management of the sand in the river.
- Beach-building flows that are greater than normal maximum operating flows are necessary to place sand on the beaches to develop camping sites.
- The transformation of the flood wave as it travels downstream from Glen Canyon Dam can be modeled, so that stage and discharge at any point and at any time along the Colorado downstream from the dam can be estimated.
- The critical reach for sand budget and beach stability is that from the Paria to the Little Colorado.
- Rapids in the Grand Canyon will continue to grow, with no mechanism to rework rapids, unless controlled floods are part of the management plan.

RECOMMENDATIONS

GCES has improved our understanding of sand movement and storage below Glen Canyon Dam. GCES sponsored the development of a more physically based analysis of the sediment transport and the eddy circulation system of the Colorado River system. GCES also advanced our understanding of the bar-building system of the Colorado River. New knowledge of changes of bar area and volume over time has given more insights into the change of the Grand Canyon river corridor and bar slope failure is better understood as a result of physically based modeling. The qualitatively rapid interchange of sand between the main channel and the eddy systems has been defined. The need for beach-building flows has been presented and is generally accepted by researchers. Several kinds of continuing research and monitoring will enhance the potential for beneficial management of sediment transport:

- Study of the rate of sand interchange between the main channel and the eddy systems that create beaches.
- Development of a mechanism for determining the initiation of beach-building flows.
- Development of quantification of the magnitude and duration of beach-building flows.
- Study of the rate at which sand is deposited on beaches during beach-building flows.
- Creation of a procedure for determining sand budgets in different parts of the canyon downstream from Glen Canyon Dam.

Less emphasis should be placed on collection of qualitative geomorphic data and more on the understanding of sediment transport processes, especially through the use of quantitative, physically based models of the system. For example, problems in using the Sediment Transport and River Simulation (STARS) water-routing model along with sediment rating curves to determine sediment transport through the Grand Canyon might have been identified earlier if more thought had been given to what was needed and how the modeling efforts would be used to meet those needs.

If the present data collection methods of the U.S. Geological Survey cannot provide adequate accuracy, support of the USGS program should be shifted to monitoring the amount of change of sand volume in the critical reaches. If the amount of sand and the change in volume in the main channel cannot be measured with accuracy, the BOR should rethink its priorities for study of sediment transport and develop an alternative method for determining the timing and magnitude of beach-building flows.

Management should coordinate the research and monitoring, even across agency lines. There should be more communication among the working groups. At the very least, each research team should know who is collecting data, where the collection will occur, and the methods that will be used. Data should be exchanged. Before going into the Grand Canyon, each team should ask whether it can collect data for other researchers.

REFERENCES

Andrews, E.D. 1991. Sediment transport in the Colorado River basin. Pp. 54-74 in Colorado River Ecology and Dam Management. Washington, D.C.: National Academy Press.

Beus, S.S., M.A. Kaplinski, J.E. Hazel, Jr., and L. Kearsley. 1994. Monitoring the Effects of Interim Flows from Glen Canyon Dam on Sand Bar Dynamics and Campsite Size in the Colorado River Corridor, Grand Canyon National Park, Arizona. Draft Final Report, U.S. Geological Survey.

Bureau of Reclamation. 1993. Operation of Glen Canyon Dam. Draft Environmental Impact Statement, U.S. Department of the Interior, Washington, D.C.

Cooley, M., B. Aldridge, and R. Euler. 1977. Effects of the catastrophic flood of December 1966, north rim area, eastern Grand Canyon, Arizona. U.S. Geological Survey Professional Paper 980.

Dawdy, D.R. 1993. Personal communication between David Dawdy and Jonathan Nelson at GCES research meeting, October 14, Boulder, Colo.

Dolan, R., and A. Howard. 1981. Geomorphology of the Colorado River in the Grand Canyon. Journal of Geology 89:269-298.

Hereford, R., H. Fairley, K. Thompson, and J. Balsom. 1991. The Effect of Regulated Flows on Erosion of Archaeologic Sites at Four Areas in Eastern Grand Canyon National Park, Arizona: A Preliminary Analysis. U.S. Geological Survey, Reston, Va.

Kieffer, S.W. 1987. The Rapids and Waves of the Grand Canyon. USGS Open-File Report 87-096, U.S. Geological Survey, Reston, Va.

National Research Council. 1987. River and Dam Management. Washington, D.C.: National Academy Press.

Randle, T.J., and E.L. Pemberton. 1987. Results and Analysis of STARS Modeling Efforts of the Colorado River in Grand Canyon. Final Report, U.S. Bureau of Reclamation, Washington, D.C.

Smith, J.D., and S. Wiele. In Press. Flow and Sediment Transport in the Colorado River Between Lake Powell and Lake Mead. Boulder, Colo.: USGS Water Resources Division.

6

Organisms and Biological Processes

INTRODUCTION

Organisms and biological processes were studied through Glen Canyon Environmental Studies (GCES) from three different perspectives: (1) as ecosystem components, (2) as resources of specific economic or recreational importance (trout), and (3) as a means of satisfying the requirements of the Endangered Species Act (humpback chub, Kanab ambersnail, bald eagle, southwestern willow flycatcher). These three perspectives, which were combined in the GCES study plan, had somewhat different objectives. Thus, the efforts of GCES were to some extent divided in three directions. In effect, ecosystem analysis, which was the unifying theme for GCES, was partially redirected by special concern over particular species. While the division of priorities is easily understandable in view of the societal value attached to trout and the legal requirements attached to endangered species, GCES was often diverted from its goal of understanding the entire system by strong focus of its resources on particular components of the system. This problem is inherent in government studies of ecological systems and can work against successful ecosystem analysis and prediction, as will be apparent from this chapter.

The biological work of GCES can be divided into four parts: (1) lake studies, (2) studies of the Colorado River between the Glen Canyon Dam and the Paria River (26 km), (3) studies of the Colorado River below the Paria to Lake Mead (450 km), and (4) studies of the riparian zone. These components are connected, of course, but each has distinctive communities and biological processes.

LAKE POWELL

Characteristics of Lake Powell

Impoundment of the Colorado River by Glen Canyon Dam has created a qualitatively new kind of water source for the Colorado River below. Characteristics of Lake Powell, together with operation of the dam, determine the temperature, suspended and dissolved solids, nutrients, and organisms passing downstream. Thus, any comprehensive assessment of the Colorado River below Glen Canyon Dam must include Lake Powell (Figure 6.1).

Lake Powell has been studied by several organizations and individuals over more than 20 years. Stanford and Ward (1991) provide an excellent interpretive summary of work completed through 1990. The Bureau of Reclamation (BOR) has supported a range of studies on Lake Powell but has taken particular interest in its effect on the mean salinity of water in the Colorado River (BOR, 1987). The National Science Foundation supported a decade-long system study of the lake that encompassed water quality, seasonal cycles, productivity, and aquatic community composition. This work is well summarized by Potter and Drake (1989). Various individual projects beginning in the late 1960s have dealt with components of the biota or specific aspects of water quality (Carothers and Brown, 1991). Thus, the starting point for GCES with respect to Lake Powell was an extensive but somewhat amorphous information base. Routine GCES studies of the reservoir did not begin until 1992, when quarterly synoptic surveys were initiated (unpublished as of 1995) and focused studies were initiated on layering of the upper water column and vertical distribution of oxygen (Marzolf, 1995).

Although the filling of Lake Powell began in 1963, the reservoir did not show strong development of stratification until about 1970 and did not fill to capacity until 1980. Thus, in a physical sense, the history of the reservoir can be divided into three intervals: 1963 to 1970, 1970 to 1980, and 1980 to the present. Optimal operation of the reservoir from the viewpoint of power production would require all water to pass through the generators at Glen Canyon Dam, and this was possible until 1980. Since 1980, however, operations entered a new era in which high runoff can require the release of water for the purpose of protecting the dam (Chapter 4). For example, use of the dam's bypass system was necessary in 1983 in order to prevent water from passing over the dam. Releases above power plant capacity are called floods, although they differ in frequency and magnitude from natural floods.

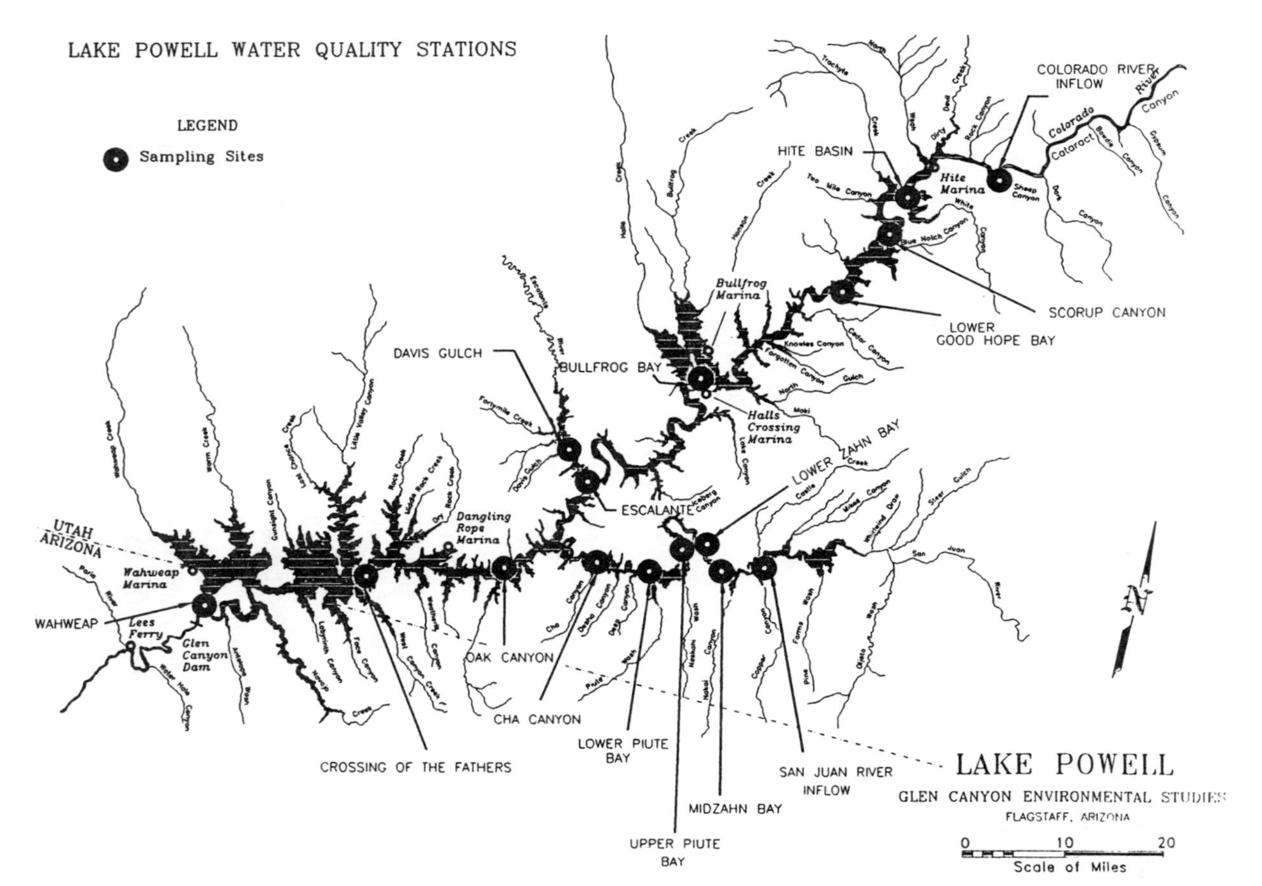

FIGURE 6.1 Lake Powell Water Quality Sampling Stations.

A succession of dry years could, of course, suppress the mean volume of the reservoir well below capacity. The reservoir might then operate for a long time with no floods, as it did between 1963 and 1983.

Although Lake Powell receives a very large amount of water from the Colorado and San Juan rivers, as well as a small amount from the Dirty Devil River, the large size of the reservoir allows a mean water retention time of approximately 2 years (Table 6.1). The residence time of the reservoir is sufficiently long to allow complete sedimentation of the inorganic material that enters it at the upper end. Lake Powell receives much sediment from its water sources (Dawdy, 1991) but, because of its great volume, has lost only a small proportion of its storage capacity through sediment accumulation.

A number of Lake Powell's features are characteristic of reservoirs in general, while others are more unusual. Like many reservoirs, Lake Powell has a dendritic shape and is much longer than broad, thus presenting the possibility of great longitudinal variations in water quality or biotic characteristics. Dendritic shape is accompanied by a high degree of shoreline development (length of shoreline). Unlike most reservoirs, however, Lake Powell has an essentially vertical shoreline around much of its perimeter. This characteristic minimizes the amount of shallow water and the potential for the development of macrophytes or other littoral communities and reduces the importance of littoral biogeochemical processes.

As would be expected for most reservoirs, the water level of Lake Powell fluctuates by several meters per year (8 m on average) and from year to year (potentially 30 m or more). Because the water withdrawal point for Lake Powell is fixed, the depth of withdrawal relative to the water surface fluctuates seasonally and also changed as the lake filled between 1963 and 1980. Recently, water has been mostly drawn from a zone 50 to 70 m below the surface, depending on season and year.

Lake Powell is divided throughout all seasons into two chemically distinct zones separated by a salinity gradient. The lower zone is consistently cold (about 7°C) and has about 50 percent more dissolved solids than the upper layer (Figure 6.2). The upper zone shows a seasonal thermal cycle that is characteristic of a warm monomictic reservoir with a substantial amount of water withdrawal. This upper zone is warmest near the surface (exceeding 25°C in the summer) and coolest near the junction with the lower layer (about 7°C).

Division of the upper zone into clearly identifiable layers that could be designated epilimnion, metalimnion, and hypolimnion would probably occur if it were not for the continuous withdrawal of large amounts of water from the

TABLE 6.1 Characteristics of Lake Powell Within 50 km of Glen Canyon Dam

Item	Amount	Source
Surface area, km^2	653	Potter and Drake (1989)
Maximum depth, m	171	Potter and Drake (1989)
Mean depth, m	51	Potter and Drake (1989)
Typical depth of water withdrawal, m	55	GCES (unpublished)
Approximate water retention time, years	2	Potter and Drake (1989)
Maximum surface temperature, °C	28	GCES (unpublished)
Minimum surface temperature, °C	7	GCES (unpublished)
Bottom temperature, °C	7	GCES (unpublished)
Suspended solids, mg/liter	7	Stanford and Ward (1991)
Total phosphorus, μg/liter	15	Gloss et al. (1980)
Dissolved phosphorus, μg/liter	10	Gloss et al. (1980)
Nitrate-N, μg/liter	300	Gloss et al. (1980)
Soluble silica (S_iO_2), mg/liter	7	Gloss et al. (1980)
Total dissolved solids, mg/liter	630	Gloss et al. (1981)
Primary production, gC/m^2/year	190	Hansmann et al. (1974)
Chlorophyll a, μg/liter	1.5	Paulson and Baker (1983) and Sollberger et al. (1989)[a]
Zooplankton abundance, individuals per liter	20	Sollberger et al. (1989)

[a]Possibly much higher; see Ayers and McKinney (1995a).

NOTE: Characteristics are for the upper 50 m of the water column unless otherwise specified.

bottom of this zone and the large additions of water at the upper end of the reservoir. Continual voluminous water exchange of this type often blurs the boundaries between thermal layers in reservoirs. In Lake Powell the distinction between metalimnion and hypolimnion is typically unclear for this reason. The summer epilimnion (mixed layer) is typically well defined, however, and the overall thermal gradient causes stability of the water column during the summer. Wind-generated mixing of the entire upper zone (50 to 70 m) occurs only during winter. Thus, the lake shows summer stratification in the upper zone, even though it has a broad thermal gradient resulting from water exchange and a lower zone that does not mix fully at any time with the upper zone.

Division of the water column into two zones is explained by seasonal fluctuations in water temperature and salinity. Water entering the reservoir during the winter is not only cold but also of high salinity relative to the water derived from snowmelt in early summer (Stanford and Ward, 1991). During the winter, water entering Lake Powell flows into the bottom zone of the res-

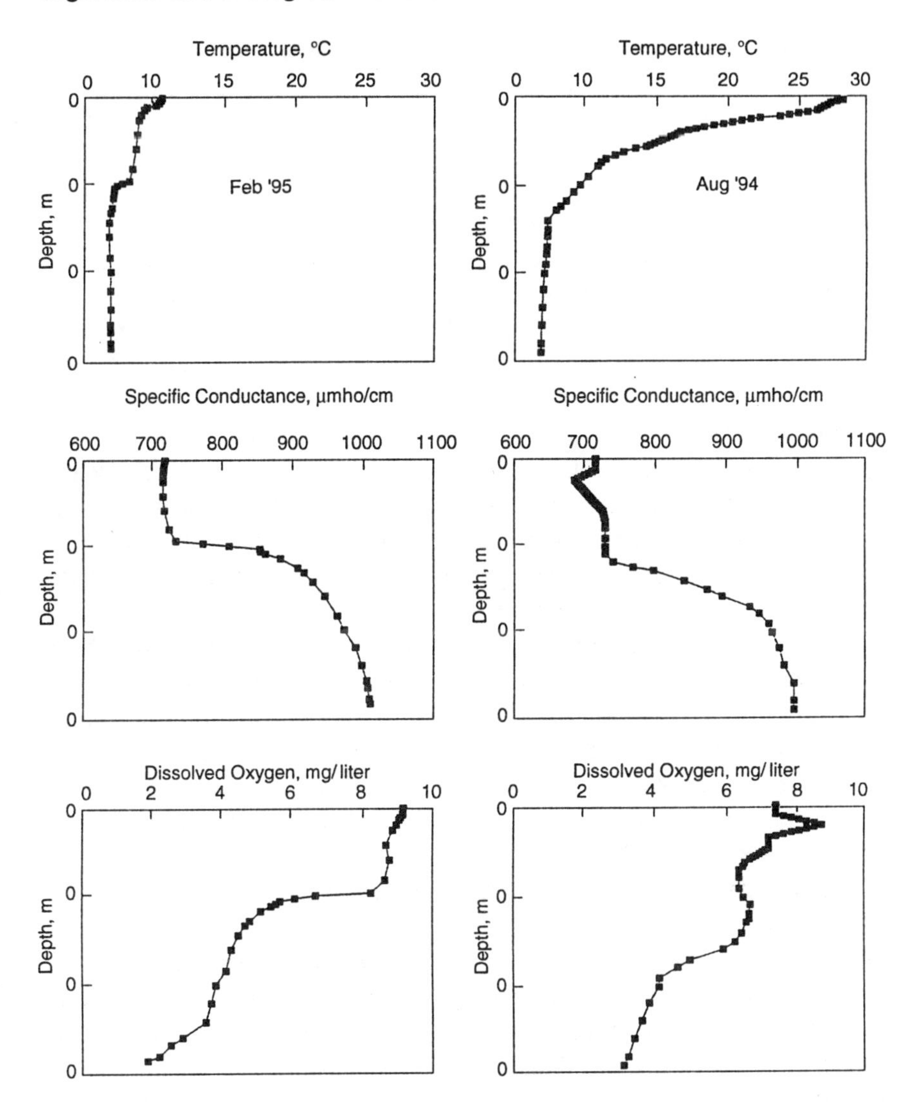

FIGURE 6.2 Profiles of temperature, conductance, and dissolved oxygen for Lake Powell near the dam at Waheap (unpublished data from the BOR). (µmho/cm = micromhos/centimeter)

ervoir and thus maintains the higher salinity of the bottom zone. During the summer, the Colorado River brings warm, less saline water derived mostly from snowmelt. This water enters the upper zone, which matches its density range. Recent studies by Marzolf (1995 and personal communication) indicate that the summer inflow enters the upper zone at a depth of 15 to 20 m and remains unmixed with the rest of the upper zone as it passes laterally all the way to the dam. Winter mixing homogenizes the upper zone, but the upper and lower zones are so different in density that they do not mix at any time of the year. Thus, separation of the zones is maintained by seasonal alternations of density for the incoming water.

Although the upper and lower zones of Lake Powell never mix fully under the influence of wind, there is a gradual interchange between the two. This interchange is quite important both to Lake Powell and to the downstream characteristics of the Colorado River. Slow exchange is caused by the addition of water to the lower layer during the winter, which increases its volume. Increase in volume of the lower layer is offset by erosion from the top of the layer under the influence of wind mixing and by withdrawal of water through the Glen Canyon Dam outlet structure. If water in the lower layer were not replaced, or were to be replaced more slowly than at present, the lower layer would become anoxic, and anoxic water could pass downstream, where it could adversely affect trout and other organisms.

Recent surveys of dissolved oxygen concentrations in Lake Powell have shown that the uppermost 10 m of the lake is often supersaturated (e.g., Figure 6.2), as would be expected for most lakes during daylight hours because of photosynthesis. There is often a metalimnetic minimum (20 to 30 m) of oxygen (e.g., Ayers and McKinney 1995a). Marzolf (1995) concludes that this minimum is caused by decomposition of organic matter in the water that enters with the spring and summer inflows at about this level.

The lowermost portions of the bottom layer of Lake Powell show substantial oxygen depletion with reference to saturation but, because of the continual replacement of water during the winter, do not become completely anoxic. Concentrations of oxygen do fall below 3 mg/liter near the bottom of the lake, however, and thus approach the limiting concentrations for the sustained presence of most kinds of aquatic life. Even minor changes in the oxygen dynamics of the reservoir could be significant. At present, however, water leaving the outlet structure is typically near saturation and experiences significant reaeration as it is discharged.

The nutrient chemistry of Lake Powell is unusual because of the large amount of suspended inorganic matter that reaches the lake from the rivers entering its upper end. Concentrations of suspended sediment in waters entering the upper end of Lake Powell have a mean of approximately 1,500 mg/liter, although concentrations can increase seasonally to as much as 10 times this amount (Stanford and Ward, 1991). Because phosphorus is readily adsorbed onto the surfaces of clay and silt particles, the total phosphorous load for Lake Powell is quite high. Although now about 20 years out of date, studies by Gloss et al. (1981) indicate that the total phosphorous loading for Lake Powell would be sufficient to maintain a hypereutrophic condition in the lake if a significant amount of the total load were in soluble form. In fact, however, over 90 percent of the phosphorous load is accounted for by phosphorus adsorbed onto inorganic particles, and this phosphorus is rapidly deposited as the reservoir sediment. Although desorption of phosphorus from sediments is possible even at the bottom of the reservoir, the interchange between the large phosphorous reserve on the bottom of the reservoir and the overlying waters is sufficiently slow that phosphorous concentrations in the water column remain low. Studies by Gloss et al. (1980), which may still be applicable, indicate that the total phosphorus of the upper water column typically ranges between 10 and 20 μg/liter within 50 to 100 km of the dam, but may be two or three times higher in the upper water column of the reservoir arms. The same studies show that dissolved phosphorus within 100 km of the dam has a mean concentration close to 10 μg/liter in the upper water column.

Concentrations of phosphorus generally decline from the upper end of the reservoir to the dam. During warm weather, dissolved phosphorus may be essentially depleted within 50 to 100 km of the dam in the upper water column. Thus, phosphorus probably limits primary production in Lake Powell near the dam, while shading caused by turbidity probably limits primary production in the upper portion of the lake (Gloss et al., 1980).

Nitrogen also reaches Lake Powell in substantial quantities, but is less associated with particulate material than phosphorus. Studies by Gloss et al. (1980, 1981) indicate concentrations of dissolved nitrogen in the upper water column of about 500 μg/liter, of which approximately half is organic N and half is nitrate N. Concentrations may be twice this high near the bottom of the lake.

Given that nitrate concentrations in Lake Powell are commonly in the vicinity of 250 μg/liter, significant nitrogen limitation of primary production seems unlikely. Gloss et al. (1980) did, however, demonstrate one instance of localized surface depletion of nitrate. Therefore, this issue is not fully resolved.

Although the most satisfactory information on nutrients is now quite old, it suggests that inorganic nitrogen passing through the outlet structure to the Colorado River is sufficient to support substantial algal growth downstream of the dam. Autotrophy just downstream of the dam seems to bear this out

(Ayers and McKinney, 1995b). Dissolved phosphorus in the outlet water would probably be in the vicinity of 10 μg/liter, which could support significant production downstream but could be removed to biologically negligible concentrations by vigorous algal growth downstream (as suggested by preliminary data from Parnell and Bennett, 1995).

Concentrations of heavy metals in Lake Powell are potentially of interest with respect to the biota of the lake as well as aquatic communities downstream. Concentrations of mercury in the water column of the lake appear to fall within the range that might inhibit growth of primary producers and may have other biological effects downstream (Graf, 1985; Blinn et al., 1977a). Selenium may also be of biological significance both within the lake and downstream (Potter and Drake, 1989). Sources of water for the reservoir contain significant amounts of radionuclides (Graf, 1985). The effects of the lake on concentrations of these substances, and their passage downstream, are undocumented.

The primary production of Lake Powell has not been estimated recently, but an estimate that is now approximately 20 years old suggests production of about 190 g of carbon per square meter per year, which is consistent with an oligotrophic or mildly mesotrophic condition (Hansmann et al., 1974; Blinn et al., 1977a). Phytoplankton abundance, as indicated by chlorophyll *a*, was between 1 and 2 μg/liters during the 1980s (Paulson and Baker, 1983; Sollberger et al., 1989). This is also consistent with general oligotrophic status, although concentrations of chlorophyll tend to be higher in the upstream arms. Algae from Lake Powell are a potential source of food for macroinvertebrates downstream. Although studies of stable isotopes have clarified some of the trophic relations downstream of the dam (Angradi, 1994), the importance of algae in supporting the food web near the dam is still unclear.

Phytoplankton composition of Lake Powell was studied by Stewart and Blinn (1976) and Blinn et al. (1977b). Primary production and biomass appear to be dominated by diatoms, and the lake features a vernal bloom that is characteristic of oligotrophic lakes. Blue-green algae probably account for only small proportions of primary production or biomass, but the information on this subject is now quite old.

Oligotrophic lakes often show significant primary production by attached algae (periphyton). The potential for periphyton growth in Lake Powell is reduced by the vertical nature of much of the shoreline, but studies by Potter and Pattison (1976) show that illuminated substrates are heavily colonized by algae and show high production per unit area.

The zooplankton community of Lake Powell was studied by Sollberger et al. (1989) and more recently by Ayers and McKinney (1995a). These studies show an unremarkable species composition that includes copepods, cladocerans, and rotifers. Densities in the upper water column were in the vicinity of 10 to 20 individuals per liter according to Sollberger et al. but were found by Ayers and McKinney (1995a) to be as high as 700 per liter (an extraordinary abundance that needs to be verified) in summer. Trout eat zooplankton that pass downstream (Angradi, 1994).

The benthic invertebrates of Lake Powell have been studied very little, but probably are dominated by chironomids (Potter and Louderbough, 1977). Fishes have been studied much more extensively (Stanford and Ward, 1991). The fish community is dominated by introduced taxa, including numerous game species. The fish community has been remarkably unstable, however, in that the dominant forage species and dominant predator species have shifted almost continually since the reservoir fish populations were first established. The lake contains numerous species that are not present downstream of the dam. Fish may survive transport through the dam, especially when some water is bypassing the turbines, but the frequency of fish passage is unknown.

Influence of Dam Operations on the Lake

The possibilities for influence of dam operations on organisms or biological processes of Lake Powell are much more limited than the reverse. The operation of a dam affects a reservoir primarily through changes in water level, mean depth, and hydraulic residence time. Because the requirements for delivery of water from Lake Powell are fixed, however, and because the reservoir is large, the flexibility for managing the water level or mean volume for any purpose other than water delivery is essentially nil. There is at present no reason to believe that any possible operating plan with the current facilities would have a distinctive influence on Lake Powell. If new facilities, such as a multiple outlet withdrawal structure or slurry pipeline, were to be installed, however, the possible effects of these structures on Lake Powell would become relevant to dam operations.

Overview of Lake Studies

Although some studies of Lake Powell were part of GCES, they were added too late and were too fragmentary to serve as an integral part of the studies. This is unfortunate, given that the water quality, temperature, and suspended organisms of Lake Powell set the initial conditions for the ecosystem downstream and may be changed either inadvertently by withdrawal at higher points in the water column in the future following a succession of dry years or by manipulation through installation of a multiple outlet withdrawal structure that would allow control of temperature for environmental purposes (BOR, 1994).

Changes in the River Caused by the Dam

Withdrawal of water from Lake Powell affects the abundance and community composition of fishes and other organisms downstream in several ways, which can be summarized as follows.

• *Hydrological changes*. The Glen Canyon Dam has drastically reduced the amount of seasonal change in the flow of the Colorado River. Dam operations also cause the water level of the river to vary on a daily basis by an amount that greatly affects the inundation of backwaters and the distribution of current velocities in the channel (Figure 6.3).

The original Colorado River was subject to surges in flow and turbidity corresponding to the flooding of tributaries associated with convectional storms. Such surges can still occur, especially below the Little Colorado River, but are reduced in frequency because of the interception of storm flows above Glen Canyon Dam by Lake Powell. Also, the suppression of floods has led to net loss of beaches and siltation of backwaters, with associated changes in habitat for organisms that require these physical features (Kearnsley et al., 1994).

• *Changes in water temperature*. The river does not become warm in the summer as it did previously (Figure 6.4). Because water is drawn from a depth that lies below the upper mixed layer of Lake Powell, water temperature is always cold near the dam (7° to 11°C). Water warms progressively downstream in midsummer (about 1°C per 30 river miles (RM); Valdez and Ryel, 1995) but reaches a summer maximum of only about 17°C near Lake Mead.

• *Changes in turbidity*. Prior to impoundment, the Colorado River was very turbid; it allowed very little light penetration. At present, the reach between Glen Canyon Dam and the Paria River is nearly always transparent.

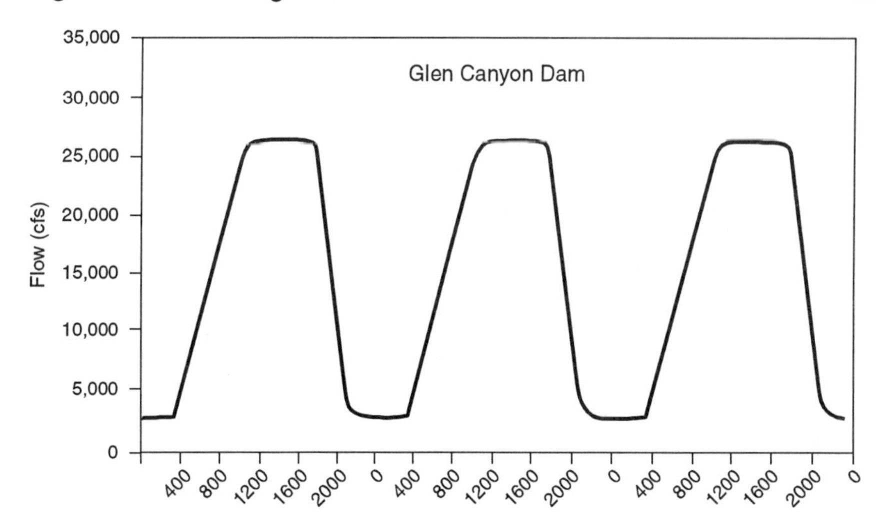

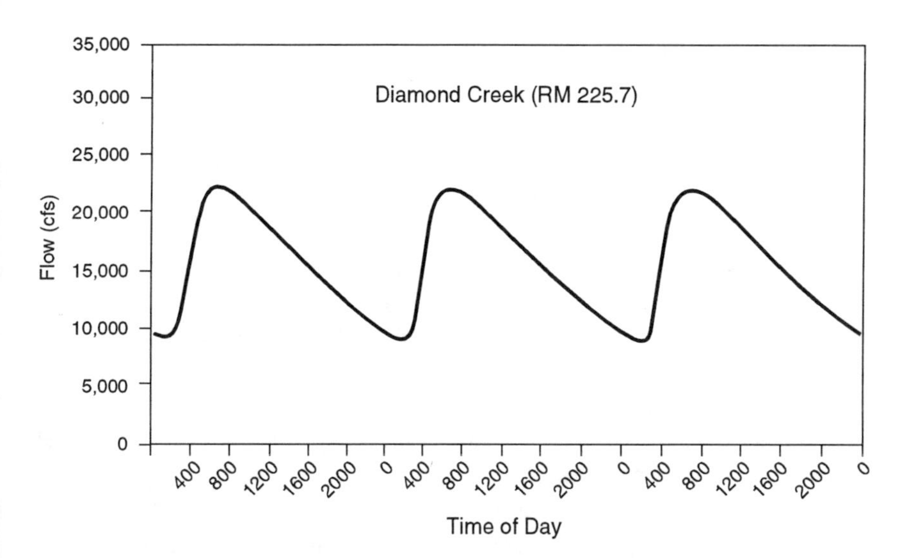

FIGURE 6.3 Discharge at two locations on the Colorado River just below Glen Canyon Dam and at Diamond Creek (both on May 10-12, 1994). SOURCE: Redrawn from Valdez and Ryel (1995).

Below the Paria, and particularly below the Little Colorado, the probability that the Colorado River will be turbid on any given day increases, but the degree and constancy of turbidity are reduced from the original state of the river.

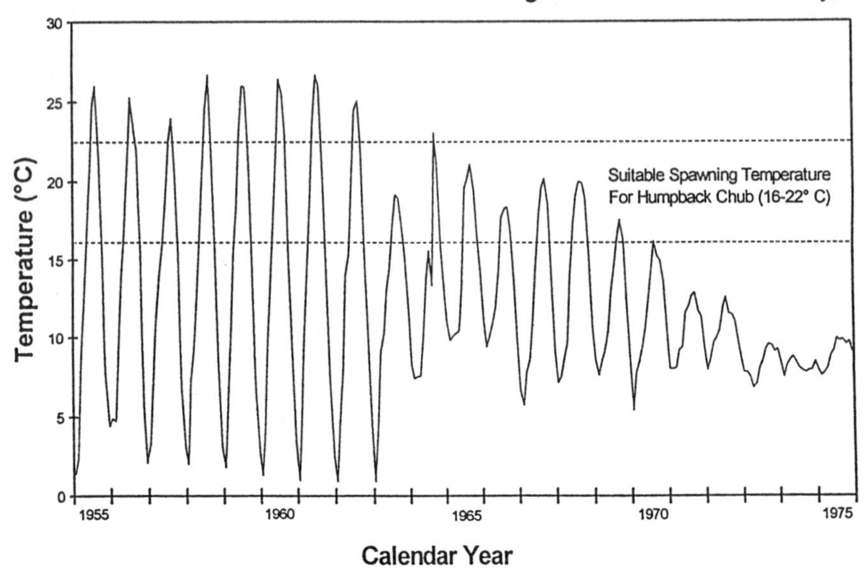

FIGURE 6.4 Mean water temperature of the Colorado River at Lee's Ferry (impoundment date was March 13, 1963). SOURCE: Redrawn from Valdez and Ryel (1995).

THE COLORADO RIVER BETWEEN GLEN CANYON DAM AND THE PARIA RIVER

The Colorado River between Glen Canyon Dam and the Paria River (26 km) lacks a sediment supply, whereas below the Paria the Colorado has a sediment supply from tributaries. Sediment supply increases stepwise with the two major tributaries (the Paria and Little Colorado) and in smaller increments with small tributaries beyond these (Chapter 5). The difference in sediment supply above and below the Paria has many ramifications that cause upper and lower sections to differ physically, chemically, and biotically.

When stripped of its sediment load by Lake Powell, the water of the Colorado River is highly transparent. The Colorado lacks high concentrations of dissolved organic carbon that might absorb light and impart color and has insufficient nutrient concentrations to generate large populations of phytoplankton in Lake Powell that would reduce transparency in the river below.

The absence of suspended sediment in the water leaving Lake Powell has led to progressive removal of sand and finer particles from the channel and sides of the Colorado River between the Glen Canyon Dam and the Paria River. Thus, the bottom of this segment of river has an abundance of coarse substrates and will show an increasing trend in this direction as the remaining stock of sand and finer particles is removed by the water from the dam. Management of flows cannot stop or reverse the loss of sediment from the upper segment because there is no source of new sediment (Andrews, 1991). Thus, any possible operating plan for the dam will lead inexorably to sediment removal, unless new facilities are installed that provide sediment through a slurry pipeline or other means. Introduction of sediment would require extensive study and justification (an environmental impact statement), and presently seems unlikely because it would probably suppress the trout population.

The condition of the Colorado River between Glen Canyon Dam and the Paria River is highly favorable for trout and highly unfavorable for native fishes, including the humpback chub (Valdez and Ryel, 1995). Replacement of fine shifting substrates with more stable gravel and cobble has favored the development and persistence of attached algae (Ayers and McKinney, 1995b), which in turn form the basis of a highly productive food web terminating in trout (see below). Aquatic productivity would not be possible without high clarity of water, which in turn is attributable to sedimentation of particulate material in Lake Powell. The presence of coarse substrates in quantity facilitates the spawning of trout, which are adapted to use such substrate, and the maintenance of significant populations of benthic insects, which are favored by coarse substrates (Blinn et al., 1994). The native fishes, which are adapted to high turbidity, fine substrates with associated eddies and backwaters, seasonally warm temperatures, and the absence of exotic carnivores such as trout, are strongly suppressed by the conditions prevailing between Glen Canyon Dam and the Paria River (Valdez and Ryel, 1995).

The biological studies of GCES have shown that the conditions of the Colorado River below Glen Canyon Dam and above Lake Mead are so inherently unsuitable for reestablishment of native fishes that a strong argument could be made for the management of this segment of the river exclusively for trout. The opposite is true downstream, as shown by the progressive disappearance of trout and a shift toward greater presence of native fishes and warmwater exotics such as carp. Thus, one possible scheme with respect to aquatic biotic resources would be to establish a framework that would allow managers to balance the welfare of the upstream resource (trout) against that of the downstream resource (native species, including the humpback chub). Unfortunately, GCES did not reach this level of integration, even though it assembled a great deal of information that could be used in creating such a framework for the rationalization of management.

Primary Producers and Their Metabolism

The most obvious photosynthetic component of the aquatic system downstream of Glen Canyon Dam is *Cladophora*, a filamentous algal species (Chlorophyta) that is widespread in inland waters. *Cladophora* is associated with running waters that offer a stable substrate (rocks), highly transparent water, and a continuous supply of inorganic nutrients. The Colorado River just below Glen Canyon Dam meets these requirements.

GCES research teams perceived the importance of *Cladophora* and the potential connection between operation of the dam and the growth of *Cladophora*. GCES produced measurements of the standing stock of *Cladophora* at various distances from the dam and of the export of *Cladophora* downstream in the form of algal clumps that break free from rocks (Blinn et al., 1994; Ayers and McKinney, 1995b). The cycling of discharge from Glen Canyon Dam on a 24-hour basis strands *Cladophora* growing in the zone of fluctuation and in this way may suppress the total standing stock of *Cladophora*. The stranding and desiccation phenomena were documented through GCES (e.g., Angradi and Kubly, 1993; Blinn et al., 1995). Moderation of daily fluctuations probably expands the standing stock of *Cladophora*.

Cladophora below Glen Canyon Dam is coated with a growth of diatoms (Blinn and Cole, 1991). Thus, *Cladophora* is significant not only for its own sake but also in providing substrate for the growth of other algae that may be more significant in the food chain. Algae other than *Cladophora* also grow directly on rocks and other substrates in the form of periphyton (Blinn et al., 1995).

Amphipods (*Gammarus*) depend on *Cladophora*, which provides the amphipods with refuge from fish predation and also supports the growth of diatoms upon which the amphipods feed. The presence and abundance of amphipods were documented by GCES. Changes in discharge are capable of dislodging amphipods or breaking away parts of the *Cladophora* (Blinn et al., 1994, 1995). Dislodged amphipods and *Cladophora* are consumed by fish (Valdez and Ryel, 1995).

As would be expected for a stony stream in the western United States, the Glen Canyon reach of the Colorado River supports benthic invertebrates that seek shelter among the stones from predation and currents. Benthic invertebrates of the Glen Canyon reach include a variety of taxa, but the fauna is especially rich in chironomids (Blinn et al., 1994, 1995), which are eaten by fish. The present fish fauna of the Glen Canyon reach reflects the influence of the dam, as do the invertebrates and algae. The fish fauna of 1963 was not, however, the same as it had been in 1850. By the time the BOR began construction of Glen Canyon Dam, the specialized native species had been joined and reduced in abundance by exotic species (Minckley, 1991) and had been adversely affected by other changes associated with land use (Miller, 1961). Thus, the dam changed the fish community but from a state that was already much altered.

The dominant fish species at present either numerically or in terms of biomass in the Glen Canyon reach is the rainbow trout (*Onchorhynchus mykiss*), which is sustained by the perennially cold water originating from the depths of Lake Powell. Reproduction of trout occurs naturally but is reinforced by stocking, which accounts for about three quarters of the catch (Maddox et al. 1987).

Trout populations had been monitored by the Arizona Department of Game and Fish (ADGF) for abundance and fishery yield over many years prior to the initiation of GCES. The routine data collection program of ADGF has produced information primarily on the numbers and sizes of fish in the population and the numbers and sizes caught by anglers. For reasons outlined by Reger et al. (1993), these fish population studies do not provide a sufficient basis for causal analysis of population dynamics. For example, a strong decline in the mean size of fish during the 1980s was of great concern to anglers but has not been satisfactorily explained.

GCES provided additional money for the study of trout, primarily through contracts between GCES and the ADGF (e.g., Maddox et al., 1987). For example, GCES supported studies of the stranding of adult trout. Stranding occurs when rapid reduction in discharge leaves trout in shallow depressions that become too warm or that lose water entirely. Although alarming to anglers, the studies show that losses probably have little effect on the total abundance of trout because the rate of loss is low in relation to population size and rate of replacement. Mortality of eggs and fry caused by fluctuations is probably much more important (Montgomery and Tinning, 1993). GCES also supported studies of relationships between *Cladophora*, amphipods, and trout. Amphipods are vulnerable to trout, especially when they are dislodged from *Cladophora*. Liebfried (1988) initially concluded that trout may derive substantial nourishment from diatoms attached to *Cladophora*, but a subsequent study by Angradi (1994) shows that trout benefit from amphipods but probably not from diatoms.

Despite substantial investment in studies of trout, GCES did not produce a comprehensive picture of the factors controlling the trout population. Growth and mortality together determine the sustainable yield from the fishery and the average size of fish caught by anglers. Neither mortality nor growth has been clearly connected to variations in the operation of Glen Canyon Dam. The effects of fluctuations in discharge on trout populations are almost as indeterminate now as they were at the beginning of GCES, despite considerable research on trout and the aquatic food web near the dam during the course of GCES.

The water temperature below Glen Canyon Dam is too low for optimal growth of trout. For this reason, installation of a multiple level outlet withdrawal structure offers the possibility of increasing trout production. Although this possibility is mentioned in a previous National Research Council review of GCES (NRC, 1987), it was not studied during Phase II of GCES.

THE COLORADO RIVER FROM THE PARIA RIVER TO LAKE MEAD

Physical Characteristics

The Paria River begins to reverse some of the effects of Glen Canyon Dam as observed in the upper segment, and the Little Colorado River reinforces this trend downstream. Downstream of these two large tributaries, as well as smaller ones, the water often carries large amounts of suspended sediment that restores, at least sporadically, the turbidity and low transparency of the original river (e.g., Melis et al., 1995). In addition, the sediment from tributaries provides a feedstock of sand for the maintenance of beaches and submerged sand deposits. Sand along the channel, in turn, creates habitat diversity both within and above the channel, and this habitat diversity supports a number of aquatic and riparian species and biological processes. The Colorado River below the Paria, while more similar to the original Colorado River than the upper reach, still differs markedly from the original Colorado. The water flowing from Glen Canyon Dam to Lake Mead even in midsummer warms only slowly in transit (ca. 1 °C for 30 miles, or 10 °C in all: Valdez and Ryel, 1995), except when it is trapped or ponded at the side of the river behind sand deposits or is warmed locally by tributaries or springs. The hydrology of the lower segment differs slightly from that of the upper segment in that the amplitude of daily change in water level declines progressively with distance from the dam. In other respects, however, the lower segment is much the same hydrologically as the upper segment. Seasonal variation of discharge is much reduced from the original condition of the river, and the river shows a daily pulse in discharge reflecting daily fluctuations in the production of hydroelectric power.

Overview of Communities

Although never measured, primary production is probably much lower in the lower reach of the Colorado River than in the upper reach because the waters downstream are frequently so turbid that algae cannot grow well (Yard et al., 1993) and because sand substrate, which is more abundant downstream, is not well suited for growth of attached algae. Thus, the Colorado River becomes progressively free of *Cladophora* downstream of the Paria. Where it is present, fixed substrate such as cobble may, however, be coated with a blue-green algal mat dominated by *Oscillatoria* (Shannon et al., 1994).

The lower reach contains smaller numbers of invertebrates than the upper reach (Blinn et al., 1994). The predominantly sandy substrate, which is poor habitat for most invertebrates, and the low abundance of *Cladophora*, which supports the large amphipod populations of the upper reach, presumably explain the decline in abundance of invertebrates below the Paria and the Little Colorado River.

The Colorado River between Glen Canyon Dam and Lake Mead was once the habitat of a highly specialized fish fauna. In the days of John Wesley Powell, the entire Colorado River Basin contained only 35 species (Valdez and Ryel, 1995). The main stem of the river in and near the Grand Canyon contained only eight species (Table 6.2), of which six were restricted to the Colorado River drainage (Minckley, 1991). A river of similar size in the eastern United States could contain over 200 fish species (e.g., Burr and Page, 1986). Thus, the original fish community of the main stem was both specialized and depauperate in number of species.

The eight taxa that originally occupied the Colorado River main stem in the Grand Canyon belong to the minnow and sucker families. Four of the five minnow species are highly unusual in morphology and life history. The Colorado River squawfish, which is the largest of the group, exceeds a length of 1 m and lives for more than 50 years (Minckley, 1991; Tyus, 1991). The three chubs are also quite large (25 to 50 cm) and long lived. Only the Colorado's speckled dace is typical of native minnows of the United States in its small size (ca. 10 cm) and short lifespan.

The chubs and squawfish of the Colorado River have unusual morphological adaptations that reflect the original conditions of the river's main stem. Their large size would have facilitated movement across or along the

TABLE 6.2 Original Fish Species of the Colorado River Between Lakes Powell and Mead

Common Name	Species Name	Status	Grand Canyon Population
Minnows (Cyprinidae)			
Colorado Squawfish	*Ptychocheilus lucius*	Endangered	Absent
Bonytail Chub	*Gila elegans*	Endangered	Absent
Humpback Chub	*Gila cypha*	Endangered	Present, localized
Roundtail Chub	*Gila robusta*	Rare	Absent
Speckled Dace	*Rhinichthys osculus*	Not rare	Present, abundant
Suckers (Catostomidae)			
Razorback Sucker	*Xyrauchen texanus*	Endangered	Present, very rare
Bluehead Sucker	*Pantosteus discolobus*	Not rare	Present
Flannelmouth Sucker	*Catostomus latipinnis*	Rare	Present

swift waters of the channel. The extended tails and dorsal humps of these taxa probably are associated with the need for high degrees of propulsion in swift waters or for maintenance of position in eddies (Valdez and Ryel, 1995). All four taxa have small eyes, presumably because visual acuity would have been of little use in the turbid waters of the original Colorado River.

Of the three species of sucker native to the Colorado River main stem along the Grand Canyon, the razorback is highly specialized in ways reminiscent of the chubs (Minckley, 1991). The other two species are less highly specialized.

Of the eight fish taxa that would have been common in the Grand Canyon region of the Colorado River at the time of the Powell expedition, only four are now self-sustaining along the river between Lakes Powell and Mead (Table 6.2). One of these, the humpback chub, is an endangered species. Changes in the fish community began well before Glen Canyon Dam was installed, but the thermal and physical effects of the dam have undoubtedly exacerbated earlier changes caused mainly by exotic species and land use or depletion of tributary flow (Miller, 1961; Minckley, 1991).

Aquatic Habitats

The Colorado River below the Paria River offers several distinctive habitat types for aquatic organisms. These can be categorized as follows: (1) open channel and bed of the river, (2) shoreline, (3) backwaters, (4) eddies, and (5) tributary junctions. The responses of aquatic organisms to these habitats are quite different; a comprehensive understanding of the biotic communities and biological processes below the Paria requires information about each. In fact, the biological component of GCES could have been organized around habitat types. Such an approach would have facilitated a transition to management-related questions if coupled to physical studies predictive of changes in the percent representation of different habitat types under a variety of operating regimes. Unfortunately, the approach to biological studies was more haphazard, and recognition of the importance of habitat types occurred more as a by-product of individual studies than as a point of departure for the entire biological component of GCES.

The open channel and bed of the Colorado River are hostile environments for most organisms because of high water velocity. Even the highly adapted humpback chub spends at most a few moments actually exposed to the full velocity of the current (Valdez and Ryel, 1995). Thus, while this habitat type is most extensive, it serves primarily as a conduit for dispersal rather than as a source of food or cover for aquatic organisms.

Even in the swiftest rivers, the shoreline is a zone of low velocity. This is particularly true where the shoreline is irregular, as is the case for some geomorphic reaches of the Colorado. The important nature of these zones is illustrated by the BioWest studies, which demonstrated the strong affinity of young of the year humpback chub for irregular shorelines (Valdez and Ryel, 1995). The extent and suitability of irregular shoreline habitat in response to differing discharge rates were not established by GCES, however, nor was the foodweb support capacity of these zones.

Backwaters include inundated off-channel areas as well as return channels associated with eddies and main-stem flow blockages behind debris. Off-channel backwaters are relatively rare in the Grand Canyon but may be important in providing nursery areas for warmwater fishes because of their warmth and low current velocities. Return channels and blockages, although typically not significantly warmer than the main channel, offer low current velocities that are critical for the humpback chub.

Backwaters were inventoried in geomorphic surveys (Holroyd, 1995) and as part of humpback chub studies. Connections between operating regimes and the area and suitability of backwaters for aquatic life were only tentatively explored in GCES, however, even though this should have been an obvious area of focus from the start.

Eddies, or recirculating zones, may provide habitat even beyond that associated with return channels. As shown by the BioWest studies, adult humpback chub are strongly associated with eddies, which probably provide an accumulation of food in a region of relatively low current velocity (Valdez and Ryel, 1995).

Tributary junctions are perhaps the most significant aquatic habitat in the entire Colorado River system below Glen Canyon Dam. They are the only persistent warmwater refugia, and they are also associated with physical habitat complexity derived from the addition of sediment and coarse debris at the tributary mouth. Habitat diversity, in the form of a wide range of current velocities, temperatures, and depths, favors diversity of aquatic life at tributary junctions.

Studies of humpback chub on the main stem demonstrate very well the critical role of the tributaries, and particularly of the Little Colorado River, in supporting the humpback chub. The main stem beyond the tributary junctions cannot support successful spawning, even though it can support adults in eddy complexes (Valdez and Ryel, 1995). Tributaries are also refugia for a wide variety of warmwater taxa. As for other habitat types, connections between operating regimes and the biological processes occurring at tributary junctions were not thoroughly worked out by GCES.

Humpback Chub

As indicated above, GCES consistently focused a great deal of its resources and attention on the humpback chub (Figure 6.5). In fact, much of the information on aquatic communities and biotic processes below the Paria River derives from the BioWest humpback chub studies (Valdez and Ryel, 1995). Special focus on the humpback chub was a by-product of the Endangered Species Act (ESA), particularly as a result of a jeopardy opinion issued by the U.S. Fish and Wildlife Service (USFWS, 1978). Because the humpback chub is a listed species, it is in effect marked as a resource of special value and was therefore an obvious candidate for special studies as part of GCES. Even more importantly, ESA requires that any changes in the management of Glen Canyon Dam be evaluated specifically for their potential effects on humpback chub and be reviewed by the U.S. Fish and Wildlife Ser-

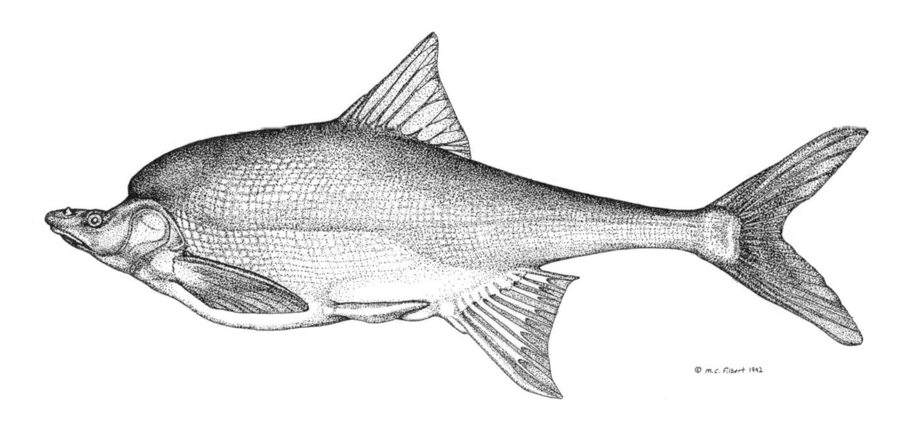

FIGURE 6.5 Humpback chub. (M. Filbert, 1993.)

vice for these effects. Finally, the potential of management practices to expand humpback chub populations is a consideration through the ESA and must be based on a thorough understanding of the requirements of the humpback chub.

The distribution and abundance of the humpback chub in the Colorado River below Glen Canyon Dam have been severely restricted by physical changes in the river. Temperature may be the most critical change, given that the humpback chub is a warmwater fish, but water clarity, availability of backwaters, seasonality of flow, and presence of exotic species are also important (Valdez and Ryel 1995). The humpback chub has survived below Glen Canyon Dam by using the Little Colorado River and other tributaries as warmwater refuges for spawning and maintenance of young fish (USFWS, 1993). Adult fish are better able to tolerate cold water and can be found in the main channel, especially in recirculation zones.

The GCES program produced extensive new information on the humpback chub below Glen Canyon Dam. GCES confirmed the concentration of the population around the Little Colorado River, quantified movements of adult chub in the main channel, and indicated the existence of small populations near eight small springs or tributaries other than the Little Colorado (Valdez and Ryel, 1995). GCES also demonstrated the presence of fry, thus providing direct evidence of reproduction, and documented wide variations in recruitment of fry from one year to another.

As acknowledged by Valdez and Ryel (1995), some important aspects of the biology of the humpback chub are not yet well documented. The requirements of very young fish, for example, are still uncertain. On the whole, however, the studies of humpback chub were successful in producing information that can be connected to management questions (Clarkson et al., 1994). The synthetic development of this information in connection with hydrological and geomorphic studies and operating regimes did not occur, however.

Other Fishes

The razorback sucker could not be studied as part of GCES because of its very low abundance (it may even be absent except in hybridized form). The other three native species still remaining are much more abundant than either the humpback chub or razorback sucker (Table 6.2). The distribution of these species is, however, focused on tributaries and is therefore probably much less extensive than it would have been originally in the Colorado River. Constantly low temperatures probably account for the restricted distribution of these warmwater taxa. Little of the GCES effort was put into studying these species, or the exotics, except as predators of humpback chub. This reflects the bias toward featured species and against the community or ecosystem perspectives.

Aside from the five native species, the lower segment of the Colorado below Glen Canyon Dam contains about 20 exotic species (Minckley, 1991). In fact, introduced species are considerably more abundant than native species either numerically or in terms of biomass. The reservoirs, and particularly Lake Mead, which provides open migratory access to the Colorado River, are a continuing source of exotics. The two reservoirs are almost entirely dominated by exotic species, and continuing changes in the fish community composition of the Colorado River may occur through entry of fishes from the reservoirs to the Colorado or its tributaries.

Fishes of the reservoirs are, to some extent, held in check by the high water velocities of the Colorado and by its low temperature. The reservoir fishes are warmwater taxa that experience some of the same life-cycle difficulties as native fishes when confronted with continuously low water temperatures. Even so, about 14 warmwater species have established populations in the warmwater refuges and have shown evidence of ability to move along the cold waters of the channel. Six coldwater salmonid species, which are not present as a substantial reproducing population in the reservoirs, are established in the Colorado River below Glen Canyon Dam, although they are less abundant in the lower reach than in the upper reach (Minckley, 1991).

Tributaries

Not only the tributary junctions but also the tributaries themselves between Glen Canyon Dam and Lake Mead are refugia for the humpback chub and other warmwater fishes. The Little Colorado River is especially important in this respect. The tributaries are flooded on an irregular basis during summer by convectional storms. Such storms can raise the discharge of the Little Colorado, for example, to 20,000 cfs or more in a very short time. These summer storms are the key means by which coarse debris, sand, and fine sediments are transported through the tributaries to the main stem. In addition, they appear to be related to the welfare of native fishes. Circumstantial evidence (Minckley and Meffe, 1987; USFWS, 1993 and BOR, unpublished) now suggests that summer floods of tributaries suppress exotic species, which are not well adapted to these conditions, and thus promote successful recruitment of native species, which are often suppressed by predation from exotics. Further study of these phenomena may illustrate the conditions under which native populations could be reinforced in tributaries other than the Little Colorado River and may highlight the importance of hydrological conditions in the Little Colorado River to the humpback chub of the Colorado River.

Both the Paria and Little Colorado rivers have been drastically altered since the days of John Wesley Powell. The riparian habitats of both of these important tributaries have been overgrazed, and this has altered their flow regimes and physical characteristics. Because of water diversion, the Little Colorado River is now essentially spring fed (Hubbs 1985) except during convectional storms, whereas previously it had significant baseflow. The welfare of the humpback chub and other native species along the Colorado River will depend on environmental changes not only in the Grand Canyon but also in tributaries of the Colorado.

Biotic Interactions

Competitive and predatory interactions involving native and exotic species might well be influenced by the operation of Glen Canyon Dam. The outcomes of such interactions are notoriously difficult to predict. GCES scarcely dealt with this subject, although biotic interactions will likely be critical in determining the final welfare of humpback chub and other native species. GCES documented the abundance and distribution of exotic fish species but primarily as an incidental matter to studies of humpback chub (Valdez and Ryel, 1995). At the community level, outcomes of operational alternatives are still highly uncertain.

Overview of Biotic Studies on the Main Stem

Studies of the aquatic foodweb by Blinn et al. (1994) and Angradi (1994), with emphasis on the area near the dam, and studies of the humpback chub in the main stem by BioWest (Valdez and Ryel, 1995) are examples of well-executed projects with specific and useful outcomes. Other studies that have not yet been finalized may prove to be of this class. The main disappointments of the GCES biotic studies are that they showed uneven coverage of the biotic resources, were largely still in progress or unreported as of the end of GCES, were weakly knit together with each other, and did not reflect a master plan for integration with hydrological and geomorphic studies that would make the essential connection to operations. However, quite a number of findings, such as those related to the habitat affinities of the humpback chub, are directly useful. In fact, a publication by Clarkson et al. (1994) that discusses the welfare of native fishes in relation to the full range of management options is a fine example of the kind of bold and broad-ranging analysis that should have marked the culmination of GCES. Interestingly, Clarkson et al. published their paper privately—presumably because federal environmental analysis is still so immature that it cannot tolerate the publication of views or conclusions that are not already reflected in the plans of management agencies. The lesson for the future is the necessity for an integrated view from the earliest planning stages to final synthesis if scientific studies are to have their fullest value in the service of management.

THE RIPARIAN ZONE

The riparian zone of a river can be loosely defined as the area that is inundated by a 100-year flood. This zone is affected by the river through scouring or deposition of sediment. Riparian zones are also influenced by their close proximity to surface water, which provides habitat and food resources that may not be present in surrounding uplands. In addition, riparian zones can typically support an abundance of vascular plants in arid climates because of the proximity of ground water to the surface. In general, riparian zones are centers of biotic diversity, especially in arid climates.

The riparian zone of the Colorado River was changed drastically by the installation of Glen Canyon Dam (Johnson, 1991). The old riparian zone extended far above the mean water level of the Colorado because of the river's typically strong spring floods, which reached discharges as high as 300,000 ft^3/second, or more than 10 times the mean discharge (Stevens and Ayers, 1993). Installation of the dam has regulated the Colorado to such an extent that the new riparian zone, if identified on the basis of high water, is much restricted; it corresponds to perhaps 90,000 cfs (the 1983 flood was just over 90,000 cfs). In fact, the riparian zone is qualitatively different now because spring flooding does not occur at all in most years, whereas strong spring flooding would have been characteristic of all years under the natural hydrological regime of the Colorado River. The original riparian zone also was affected by summer floods (convectional storm floods) and may still be under present circumstances below the Little Colorado River (e.g., Stevens and Ayers, 1993), but not between the Little Colorado and Glen Canyon Dam. Riparian characteristics associated with proximity to surface water and with the availability of ground water near the surface are the basis for maintenance of characteristic vegetation and associated riparian species below Glen Canyon Dam. In fact, ground water supply to the riparian zone is more constant now than before 1963 because the operation of the dam has stabilized the distribution of riparian ground water. This in part explains the general vegetative enrichment of the riparian zone following closure of the dam (Johnson, 1991).

The upper end of the riparian zone is characterized by hackberry, acacia, mesquite, and other species that were established by and tolerant of occasional flooding well above the mean annual flood. These species do not have access to phreatic water because they are too far from the river to reach the water table. Remaining individuals in this upper zone are old and probably will not be replaced because of the suppression of flooding. Flood-intolerant plants appear to be colonizing this zone now (Anderson and Ruffner, 1987). Restoration of the conditions leading to the establishment of this portion of the riparian zone (old high-water zone) are not within the scope of any feasible operation for Glen Canyon Dam.

The lower portion of the riparian zone contains native species, such as willow, and exotic species, such as tamarisk. In many locations, tamarisk forms a riparian forest close to the water. Tamarisk now provides habitat for both native and exotic animals and figures importantly in the biology of the riparian zone. The woody taxa in general are dependent on sandy substrate deposited by the river and on phreatic water, which is also maintained by the river. The abundance of tamarisk and the general proliferation of woody species are attributable to the dam in the sense that the annual floods of the original river would have removed such vegetation, just as the 1983 flood did (Stevens and Waring, 1985; Pucherelli, 1988). In addition, fluvial marshes have developed in places where they would have been swept away by the natural hydrological regime of the Colorado (Stevens et al., 1994).

GCES produced information on the composition of riparian vegetation (Stevens and Ayers, 1994) and lists of resident species of birds (Sogge et al., 1994), with particular attention to endangered species (Sogge and Tibbitts, 1994a,b). In fact, the major focus of GCES in the riparian zone was inventory. Functional relationships of organisms in the riparian zone and their responses to the operation of the Glen Canyon Dam were also studied by GCES but less completely and with a more variable outcome. For example, Stevens and Kline (1991) concluded that large daily fluctuations may be detrimental to waterfowl, while Stevens and Ayers (1993b) showed that moderation of daily fluctuations during the period of interim flows probably had no significant adverse effects on the riparian zone. Controlled or uncontrolled floods are of particular relevance to the riparian zone, but the forecasting of effects for such events is weak at present. For the future, responses of the riparian zone to operation of the dam should be an important consideration of adaptive management, but the scientific basis for this should be strengthened.

OUTCOMES OF BIOLOGICAL STUDIES

GCES provided much new information on the biotic resources of the Colorado River below Glen Canyon Dam. The abundances and distributions of many kinds of organisms were quantified satisfactorily for the first time. In addition, several kinds of functional relationships, such as the dependence of rainbow trout on amphipods and *Cladophora* near Glen Canyon Dam, were documented. Many functional relationships, however, were not explored satisfactorily or were not explored at all. Some of these relationships are critically connected to management options, but studies of them were not initiated or did not come to completion in a way that would be useful to management. Finally, the integration of biotic components with each other, and joint consideration of biological and physical aspects of the environment, particularly involving sediment dynamics, remained largely undeveloped as of the end of GCES.

A few examples will illustrate some of the strengths and weaknesses of the biological component of GCES studies. Present management options include a range of potential discharge manipulations. Managers might ask for a list of fishes and other species that might be affected by such manipulations. GCES could easily provide a comprehensive list, along with relative abundances, probable habitat requirements, and general distribution downstream of Glen Canyon Dam. Management might, however, also ask for a forecast of the effects on particular species. For example, how would greater stability of discharge affect species and communities downstream? Here GCES might prove to be only partially satisfactory. For example, GCES showed that large rapid fluctuations in discharge interfere with spawning of trout near Glen Canyon Dam and strand some adult fish. At the same time, however, such variations appear to dislodge food items that in this way become vulnerable to trout. What is the relative importance of these two effects on the trout population? GCES does not provide an answer, although clearly management needs at least a qualitative answer to such questions. In general, the success of GCES was greatest in the area of survey and inventory, impressive in the analysis of some (e.g., native fishes) but not all system components, and disappointing in the area of ecosystem analysis.

REFERENCES

Anderson, L.S. and G.A. Ruffner. 1987. Effects of Post-Glen Canyon Flow Regime on the Old High Water Line Plant Community Along the Colorado River in the Grand Canyon. Glen Canyon Environmental Studies Technical Report, Bureau of Reclamation, Salt Lake City, Utah.

Andrews, E.D. 1991. Sediment transport in the Colorado River Basin. Pp. 54-74 in Colorado River Ecology and Dam Management. Washington, D.C.: National Academy Press.

Angradi, T.R. 1994. Trophic linkages in the lower Colorado River: multiple stable isotope evidence. Journal of the North American Benthological Society 13:479-493.

Angradi, T.R., and D.M. Kubly. 1993. Effects of atmospheric exposure on chlorophyll *a*, biomass and productivity of the epilithon of a tailwater reservoir. Regulated Rivers: Research and Management 8:345-358.

Ayers, A.D. and T. McKinney. 1995a. Water Chemistry and Zooplankton in the Lake Powell Forebay and the Glen Canyon Dam Tailwater. Draft final report, Arizona Game and Fish Department, Phoenix.

Ayers, A.D., and T. McKinney. 1995b. Effects of Different Flow Regimes on Periphyton Standing Crop and Organic Matter and Nutrient Loading Rates for the Glen Canyon Dam Tailwater to Lee's Ferry. Draft final report. Arizona Game and Fish Department, Phoenix.

Blinn, D.W. and G.A. Cole. 1991. Algal and invertebrate biota in the Colorado River: comparison of pre- and post-dam conditions. Pp. 102-123 in Colorado River Ecology and Dam Management. Washington, D.C.: National Academy Press.

Blinn, D.W., T. Tompkins, and L. Zaleski. 1977a. Mercury inhibition on primary productivity using large volume plastic chambers in situ. Journal Phycology 13:58-61.

Blinn, D.W., T. Tompkins, and A.J. Stewart. 1977b. Seasonal light characteristics for a newly formed reservoir in southwestern USA. Hydrobiologia 51:77-84.

Blinn, D.W., L.E. Stevens, and J.P. Shannon. 1994. Interim Flow Effect from Glen Canyon Dam on the Aquatic Food Base in the Colorado River in the Grand Canyon National Park, Arizona. Cooperative Study Agreement CA824-8-0002.

Blinn, D.W., J.P. Shannon, L.E. Stevens, and J.P. Carder. 1995. Consequences of fluctuating discharge for lotic communities. Journal of the North American Benthological Society 14:233-248.

Bureau of Reclamation. 1987. Quality of Water, Colorado River Basin. Progress Report No. 13., U.S. Department of the Interior, Washington, D.C.

Bureau of Reclamation. 1994. Glen Canyon Dam Discharge Temperature Control. Draft Appraisal Report, Bureau of Reclamation, Washington, D.C.

Burr, B.E.M., and L.M. Page. 1986. Zoogeography of the fishes of the lower Ohio-upper Mississippi basin. Pp. 287-324 in The Zoogeography of North American Freshwater Fishes, C.H. Hocutt and E.O. Wiley, eds. New York: John Wiley.

Carothers, S., and B. Brown. 1991. The Colorado River Through Grand Canyon: Natural History and Human Change. Tucson: University of Arizona Press.

Carothers, S., and C. Minckley. 1981. A survey of the fishes, aquatic invertebrates and aquatic plants of the Colorado River and selected tributaries from Lee's Ferry to Separation Rapids. Prepared by Museum of Northern Arizona for Water and Power Resources Service, Lower Colorado Region, Boulder City, Nev.

Clarkson, R.W., O.T. Gorman, D.M. Kubly, P.C. Marsh, and R.A. Valdez. 1994. Management of Discharge, Temperature, and Sediment in Grand Canyon for Native Fishes. Report privately published by the authors.

Dawdy, D.R. 1991. Hydrology of Glen Canyon and the Grand Canyon. Pp. 40-53 in Colorado River Ecology and Dam Management. Washington, D.C.: National Academy Press.

Gloss, S.P., L.M. Mayer, and D.E. Kidd. 1980. Advective control of nutrient dynamics in the epilimnion of a large reservoir. Limnology Oceanography. 25:219-228.

Gloss, S.P., R.C. Reynolds, Jr., L.M. Mayer, and D.E. Kidd. 1981. Reservoir influences on salinity and nutrient fluxes in the arid Colorado River Basin. Pp. 1618-1629 in H.G. Stefan, ed. Proceedings of the Symposium on River Surface Water Impoundments. New York: American Society of Civil Engineers.

Graf, W.L. 1985. The Colorado River: Instability and Basin Management. Resource Publication in Geography, American Society of Geographers, Washington, D.C.

Hansmann, E.W., D.E. Kidd, and E. Gilbert. 1974. Man's impact on a newly formed reservoir. Hydrobiologia 45:185-197.

Holroyd, E.W. 1995. Thermal infrared (FLIR) studies in eastern Grand Canyon. Technical Memorandum #8260-95-11. Bureau of Reclamation, Washington, D.C.

Hubbs, C. 1995. Springs and Spring Runs as Unique Aquatic Ecosystems. Copeia 1995:989-991.

Johnson, R.R. 1991. Historic changes in vegetation along the Colorado River in the Grand Canyon. Pp. 178-206 in Colorado River Ecology and Dam Management. Washington, D.C.: National Academy Press.

Kearnsley, L., J. Schmidt, and K. Warren. 1994. Effects of Glen Canyon Dam on Colorado River Sand Deposits Used as Campgrounds in Grand Canyon National Park, USA. Regulated Rivers 9:137-149.

Liebfried, W.C. 1988. The utilization of *Cladophora glomerata* and Epiphitic Diatoms as a Food Source by Rainbow Trout in the Colorado River Below Glen Canyon Dam, Arizona. Master's thesis, Northern Arizona University, Flagstaff.

Maddox, H.R., D.M. Kubly, J.C. Devos, W.R. Persons, R. Staedlcke, and R.L. Wright. 1987. The Effects of Varied Flow Regimes on Aquatic Resources of Glen and Grand Canyons. Arizona Game and Fish Department.

Marzolf, R. 1995. Metalimnetic oxygen depletion in Lake Powell: a consequence of the inflow of the annual snowmelt water mass. Paper presented at national ASLO meeting, June 11-15, Reno, Nev.

Melis, T.S., W.M. Phillips, R.H. Webb, and D.J. Bills. 1995. When the Blue-Green Waters Turn Red: A History of Flooding in Havasu Creek, Arizona. USGS Water-Resources Investigation Report, U.S. Geological Survey, Reston, Va.

Miller, R.R. 1961. Man and the changing fish fauna of the American southwest. Michigan Academy of Science, Arts and Letters 46:365-404.

Minckley, W.L. 1991. Native fishes of the Grand Canyon region: an obituary? Pp. 124-177 in Colorado River Ecology and Dam Management. Washington, D.C. National Academy Press.

Minckley, W.L. and G.K. Meffe. 1987. Differential selection by flooding on stream-fish communities in the arid American Southwest. Pp. 93-104 in D.C. Heins and W.J. Matthews (eds), Evolutionary and Community Ecology in North American Stream Fishes. University of Oklahoma Press.

Minckley, W.L., P.C. Marsh, J.E. Brooks, J.E. Johnson, and B.L. Jensen. 1991. Management toward recovery of the razorback sucker. Pp. 303-358 in Battle Against Extinction: Native Fish Management in the American West, W.L. Minckley and J.E. Deacon, eds. Tucson: University of Arizona Press.

Montgomery, W.L., and K. Tinning. 1993. Impacts of fluctuating water levels on eggs and fry of rainbow trout in the Colorado River below Glen Canyon Dam, Arizona. Unpublished manuscript.

National Research Council. 1987. River and Dam Management: A Review of the Bureau of Reclamation's Glen Canyon Environmental Studies. Washington, D.C. National Academy Press.

Parnell, R.A., and J.B. Bennett. 1995. Influence of geochemical processes on nutrient spiralling within the recirculation zones of the Colorado River and the Grand Canyon. Quarterly Report, April. National Park Service Cooperative Agreement CA8000-8-0002, National Park Service, Washington, D.C.

Paulson, L.J. and J.R. Baker. 1983. Limnology in reservoirs on the Colorado River. Technical Compl. Report OWRT-B-121-NEV-1, Nevada Water Resources Research Center, Las Vegas.

Potter, L.D., and C. Drake. 1989. Lake Powell: Virgin Flow to Dynamo. Albuquerque: University of New Mexico Press.

Potter, L.D., and E.T. Louderbough. 1977. Macroinvertebrates and diatoms on submerged bottom substrates, Lake Powell. Lake Powell Research Project Bulletin 37, Institute of Geophysics and Planetary Physics, University of California, Los Angeles.

Potter, L.D., and N.B. Pattison. 1976. Shoreline ecology of Lake Powell. Lake Powell Research Project Bulletin 29, Institute of Geophysics and Planetary Physics, University of California, Los Angeles.

Pucherelli, M.J. 1988. Evaluation of riparian vegetation trends in the Grand Canyon using multitemporal remote sensing techniques. Pp. 217-228 in Glen Canyon Environmental Studies: Executive Summaries of Technical Reports. U.S. Department of the Interior, Bureau of Reclamation, Salt Lake City.

Reger, S., C. Benedict, and D. Wayne. 1993. Statewide Fisheries Investigations: Survey of Aquatic Resources, Colorado River, Lee's Ferry. Draft Fish Management Report 1989-1993, Federal Aid Project F-7-M-36.

Shannon, J.P., D.W. Blinn, and L.E. Stevens. 1994. Trophic interactions and benthic animal community structure in the Colorado River, Arizona, U.S.A. Freshwater Biology 31:213-220.

Sogge, M.K., and T.J. Tibbitts. 1994a. Wintering Bald Eagles in the Grand Canyon: 1993-1994. Summary Report, National Biological Survey Colorado Plateau Research Station/Northern Arizona University and U.S. Fish and Wildlife Service, Phoenix.

Sogge, M.K., and T.J. Tibbitts. 1994b. Distribution and Status of the Southwestern Willow Flycatcher Along the Colorado River in the Grand Canyon—1994. Summary Report, National Biological Survey Colorado Plateau Research Station/Northern Arizona University and U.S. Fish and Wildlife Service, Phoenix.

Sogge, M.K., D. Felley, P. Hodgetts, and H. Yard. 1994. Grand Canyon Avian Community Monitoring: 1993-94 Progress Report. National Biological Survey Colorado Plateau Research Station Report.

Sollberger, J.P., P.D. Vaux, and L.J. Paulson. 1989. Investigation of Vertical and Seasonal Distribution, Abundance and Size Structure of Zooplankton in Lake Powell. Lake Mead Limnological Research Center, University of Nevada, Las Vegas.

Stanford, J.A., and J.V. Ward. 1991. Limnology of Lake Powell and the chemistry of the Colorado River. Pp. 75-101 in Colorado River Ecology and Dam Management. Washington, D.C.: National Academy Press.

Stevens, L.E., and T.J. Ayers. 1993a. The Impacts of Glen Canyon Dam on Riparian Vegetation and Soil Stability in the Colorado River Corridor, Grand Canyon, Arizona. 1992 Final Administrative Report, National Park Service Cooperative Studies Unit, Northern Arizona University, Flagstaff, Ariz.

Stevens, L.E., and T.J. Ayers. 1993b. The Effects of Interim Flows from Glen Canyon Dam on Riparian Vegetation Along the Colorado River in Grand Canyon National Park, Arizona. Draft 1992 Annual Report, NPS Cooperative Work Order No. CA 8021-8-0002, National Park Service, Grand Canyon, Ariz.

Stevens, L.E., and T.J. Ayers. 1994. The Effects of Interim Flows from Glen Canyon Dam on Riparian Vegetation Along the Colorado River in Grand Canyon National Park, Arizona. Draft 1994 Annual Report, NPS Cooperative Work Order No. CA 8021-8-0002, National Park Service, Grand Canyon, Ariz.

Stevens, L.E., and N. Kline. 1991. 1991 Aquatic and semi-aquatic avifauna in the Colorado River corridor in the Grand Canyon, Arizona. Draft report.

Stevens, L.E., and G. Waring. 1985. The effects of prolonged flooding on the riparian plant community in Grand Canyon. Pp. 81-86 in Riparian Ecosystems and Their Management: Reconciling Conflicting Uses. USDA Forest Service General Technical Report RM-120. Rocky Mountain Forest and Range Experiment Station, USDA Forest Service, Ft. Collins, CO.

Stevens, L.E., J.C. Schmidt, T.J. Ayers, and B.T. Brown. 1994. Fluvial marsh development along the dam-regulated Colorado River in the Grand Canyon, Arizona. Unpublished manuscript.

Stewart, A.J., and D.W. Blinn. 1976. Studies on Lake Powell, U.S.A.: environmental factors influencing phytoplankton succession in a high desert warm monomictic lake. Arch. Hydrobiology 78:139-164.

Tyus, H.M. 1991. Ecology and management of Colorado squawfish. Pp. 379-404 in Battle Against Extinction: Native Fish Management in the American West, W.L. Minckley and J.E. Deacon, eds. Tucson: University of Arizona Press.

U.S. Fish and Wildlife Service. 1978. Biological Opinion on the Effects of Glen Canyon Dam on the Colorado River as It Affects Endangered Species. Memorandum issued from Albuquerque, N.Mex.

U.S. Fish and Wildlife Service. 1993. Habitat Use by Humpback Chub, *Gila cypha*, in the Little Colorado River and Other Tributaries of the Colorado River.

Valdez, R.A., and R.J. Ryel. 1995. Life History and Ecology of the Humpback Chub (*Gila cypha*) in the Colorado River, Grand Canyon, Arizona. Draft Final Report, BioWest, Inc., Logan, Utah.

Yard, N.D., G.A. Hayden, and W.S. Vernieu. 1993. Photosynthetically Available Radiation (PAR) in the Colorado River: Glen and Grand Canyons. Glen Canyon Environmental Studies Technical Report.

7

Recreation and Nonuse Values

THE ROLE OF ECONOMIC VALUES IN GCES AND THE EIS

Much of the research under the Glen Canyon Environmental Studies (GCES) was physical and biological. Economic research also became an integral component of the research program. Economic research had three foci. First, economic tools were used to quantify, in monetary terms, the effects of dam operations on the quality of river recreation. Second, economic tools were used in analyzing the effects of dam operations on values held by those who do not directly use resources affected by dam operations (i.e., nonuse values). Third, economic analysis was used in quantifying the linkage between operating criteria and value of the electricity generated at the dam.

It is important to keep in mind that the economic studies focus on valuing the effects of alternative dam operations, not on the river as a whole. This chapter addresses the use of economic methods to document changes in the value of recreational opportunities associated with changes in dam operations. It also explains the concept of nonuse values, their relevance to dam operations, and the results of nonuse value studies. Chapter 9 summarizes and evaluates GCES work on the economics of power generation at Glen Canyon Dam.

Use and Nonuse Values Defined

Recreation and power values are *use values* because they stem from the direct use of river resources to produce electrical and recreational benefits.

Policymakers, economists, and the public question whether the economic values of environmental resources should be limited to use values (HBRS, 1991, Harpman et al., 1995). For example, those who have not visited the Grand Canyon may place an economic value on the preservation of its resources for future generations or their own option to use the canyon in the future. Such values are often called *nonuse values*. They are motivated by value attached to the continued existence or preservation of a resource or the resource's for future generations (Chapter 3). Nonuse values are not held only by "nonusers." Visitors to the river corridor below the dam may hold nonuse value in addition to use value. Environmental economists have developed a theory of *total value*, which consists of use and nonuse values (HBRS, 1991, Harpman et al., 1995).

Questions about the effects of dam operations on the total value of the resources downstream from Glen Canyon Dam are appropriate because federal law requires, as part of the environmental impact statement (EIS) process, consideration of the economic implications of alternatives.

The economic theory and empirical measurement techniques relevant to nonuse values in resource valuation studies have evolved rapidly during the past decade (HBRS, 1991, Harpman et al, 1995). As a result, nonuse values have been included in a variety of policy analyses for which changes in the quality or availability of natural resources are an issue. Perhaps the most important example is the rules for assessing damages to natural resources from spills of oil and toxins under the Comprehensive Environmental Response, Compensation, and Liability Act and the Clean Water Act (U.S. Department of the Interior, 1991). A U.S. Court of Appeals decision in 1989 strengthened the role of nonuse values in such cases, and nonuse values were important in arriving at a negotiated settlement on liability for the *Exxon Valdez* oil spill. More recently, the National Oceanic and Atmospheric Administration convened a blue-ribbon panel that evaluated the validity of methods for measuring nonuse values and developed guidelines for measuring nonuse values in natural resource damage assessment (NOAA, 1993). In addition, several federal agencies are writing administrative rules for the measurement and application of nonuse values to public policy processes.

Measurement of Nonuse Values

While the validity of nonuse values is well established in theory, such values cannot influence policy decisions unless they can be measured accurately. Measurement of nonuse values relies on the contingent valuation method (CVM), which quantifies willingness to pay. There has been substantial debate among economists and other social scientists over the quantification of willingness to pay. Although contingent valuation continues to be controversial, there is a growing body of evidence that supports its practical usefulness (Harpman et al., 1995). Contingent valuation is routinely applied with confidence to estimates of use values, and early work on nonuse values is encouraging.

Whether nonuse values can be measured with sufficient accuracy to meet high scientific standards is a question still widely discussed among policy analysts and economists. There is, however, a theoretical economic framework sufficient to form a foundation for their use in the GCES. The literature on CVM indicates that accuracy is sufficient to make quantification of nonuse value useful in understanding the balance of values at stake in managing Glen Canyon Dam. This is particularly true given all that can be learned in the nonuse valuation process regarding public views of the resource issues being addressed under GCES. To neglect total values in favor of more narrowly defined use values would be to leave a major gap in the economic studies under GCES and in the Glen Canyon Dam EIS. This would be unjustifiable given that nonuse values can be estimated.

OVERVIEW OF RECREATIONAL USES

Recreation is an important use of the Colorado River below Glen Canyon Dam. Each year over 20,000 anglers, 33,000 day-trip rafters, and 15,000 to 20,000 white-water boaters use this section of the river. The GCES examined recreational use patterns and values in considerable detail and focused on those types of recreation most likely to be affected by changes in the dam's operations.

The 15-mile segment of the Colorado River immediately below Glen Canyon Dam is located in the Glen Canyon National Recreation Area. It is used by a variety of recreationists, including fishermen, boaters, day rafters, campers, bird watchers, and hikers. Below the Glen Canyon reach the Colorado River flows through Marble and Grand canyons for 277 miles, including over 160 recognized rapids. Some of the world's most challenging and exciting white waters occur here. Below the Grand Canyon, Hoover Dam holds back the Colorado River to form Lake Mead, which is one of the largest reservoirs in the western United States. The dam's operation affects the experience of recreationists using the Colorado River in Glen Canyon and the Grand Canyon.

In 1987 a study of river-based recreation between Lakes Powell and Mead was completed by Bishop et al. The goals of the study were to document the quantity and pattern of river-based recreational use, to identify factors having a significant effect on the net economic value of recreational use, and to estimate the net economic value of river-based recreation. The authors identified four major categories of river-based recreational use: (1) day (scenic) rafting in Glen Canyon, (2) angling in Glen Canyon, (3) commercial white-water boating in Grand Canyon, and (4) private white-water boating in Grand Canyon.

Bishop et al.'s early survey work (Bishop et al., 1987) involving anglers and boaters determined that the value of angling and white-water boating is affected by river stage and daily fluctuations but that day rafters are not particularly sensitive to these aspects of dam operations. Consequently, the economic effects of operational alternatives on day rafters are negligible.

Fishing in Glen Canyon

The Glen Canyon trout fishery is a by-product of Glen Canyon Dam. Discharge from the dam is colder, carries less silt, and is more stable on an annual basis than prior to construction of the dam. This altered environment supports a good trout fishery. The Arizona Department of Game and Fish (ADGF) stocks up to 100,000 rainbow trout in some years; in more recent years, brook trout and cutthroat trout also have been stocked. Surveys of Arizona anglers conducted by ADGF indicate that trout are the most desired sport fish in the state, but preferences among trout species and between native and stocked trout have not been well documented, as pointed out in 1987 by the NRC committee (NRC, 1987).

The introduced trout have created an important fishery that is considered to be of high quality. Glen Canyon is one of only two blue-ribbon stream fisheries in Arizona. Each year more than 19,000 anglers fish for rainbow trout in the 15-mile reach below the dam.

Bishop et al.'s (1987) study also revealed that the attributes most strongly affecting the Glen Canyon fishing experience are the site and number of fish the respondent expected to catch. Fishing success is believed to be related to flow in two ways. Rising water may improve fishing as fish begin to feed on invertebrates that are dislodged in this way. In addition, flows of 10,000 cubic feet per second (cfs) and less provide gravel and rock bars for fishing and some room for bank fishing between the water's edge and shore vegetation. Low flows limit boaters ability to get upstream.

Fishing in Grand Canyon

Fishing in Grand Canyon is likely an activity incidental to white-water boating or backpacking, except in some side canyons around Marble Canyon. The National Park Service (NPS) controls access to these wild trout fisheries by issuing back-country and river permits. Commercial river companies are not allowed to offer trips that are primarily for fishing in Grand Canyon, even though fishing is allowed as an incidental activity on river trips.

Day Rafting

In 1991 more than 33,000 visitors took half-day tours of the Glen Canyon reach. Bishop et al. (1987) found that the only flow-sensitive attribute of a Glen Canyon day-raft trip may be from where the trip originates. At low-to-moderate flow levels (generally less than 29,500 cfs), the 20-person tours depart from a dock near Glen Canyon Dam and float or motor downstream to Lee's Ferry. When releases are above 29,500 cfs and outlet works are in use, departure from the base of the dam is unsafe due to the volume and turbulence of the water. In these cases, rafts normally depart from Lee's Ferry carrying fewer people (10) and motor part way upstream before floating back downstream.

White-Water Boating

White-water boating in Grand Canyon is a major industry; 21 companies have permits to conduct commercial raft trips in the park. Also, the Hualapai Tribe conducts river trips from Diamond Creek to Lake Mead. Each year 15,000 to 20,000 commercial and private boaters annually run the river. The number of user-days is restricted to 115,500 for commercial trips and 54,450 for private parties. White-water boating use is limited to 166 visitors per day during the primary season (May 1 through September 30). These limitations were designed to maintain boating safety, reduce crowding on the river, and minimize damage to riparian resources. The regulations preclude any increases in use of the river for white-water boating. Motorized trips are allowed to launch from mid-December through mid-September. Oar-powered craft can be used throughout the year. The size of private groups averages less than the limit of 16, while commercial group size usually is 36 people. The lower gorge, which begins at Diamond Creek, is used by the Hualapai Tribe concession as well as by other commercial and private rafters. Most commercial and private raft trips take place between May and October.

Bishop et al. (1987) asked white-water boaters to identify the attributes that contribute most to an excellent Grand Canyon white-water trip. Of the attributes listed by at least 15 percent of all respondents, four are potentially affected by dam operations: time for layovers and stops at specific sites, rapids, wilderness experience, and not feeling crowded.

River trips make planned stops at attractions located along the tributaries and side canyons and also include scheduled short or extended day hikes. These stops are important attributes of white-water trips. During low flows, both commercial and private trip passengers may miss one or more sites because of the additional time needed on the river to maintain a trip schedule.

Rapids are also important attributes of white-water boating trips (Bishop et al., 1987). Rapids are flow related because a number of small-to-medium rapids become less exciting to run at high flows, while larger rapids generally become more exciting to run at high flow. Also, guides and trip leaders are more likely to have passengers walk around major rapids at the highest flows (above 35,000 cfs). At low flows (5,000 cfs or less), it often becomes necessary to either walk passengers around some rapids or to wait for higher water.

One of the attributes of an excellent river trip most often identified by river runners is a wilderness experience. Enjoying a wilderness experience is more important to private rafters and oar trip passengers and less important to commercial and motor passengers. Most river runners are aware of wide daily fluctuations, and most feel that the fluctuations make the setting for the trip seem less natural (Bishop et al., 1987).

White-water boaters may feel most crowded at high flows because the number and size of beaches for camping are significantly reduced. In addition, during daily fluctuations in flows, boaters may congregate above rapids as they wait for the water level to rise. Flows affect the usable area of a camping beach. The rise and fall of water levels that result from fluctuating discharges (Chapter 4) inundate portions of beaches, strand boats, and influence the character of the setting. Daily fluctuations influence campsite selection in that many river runners will not choose a campsite that does not offer protection against changes in water level (Bishop et al., 1987). An average of 35 percent of potential campsite area is inundated when releases increase from 5,000 to 25,000 cfs.

Bishop et al. (1987) asked white-water boaters and commercial white-water guides to provide reports on the quality of their Grand Canyon white-water trips. Both the guides and the passengers reported the highest quality for trips during periods of constant flows in the range of 25,000 to 30,000 cfs. For a variety of reasons, however, it is impractical to release 30,000 cfs over long periods of time to increase the quality of white-water trips (Chapter 4).

River levels affect accident rates; flood flows and low flows are believed to be the most hazardous. Fluctuating flows are not considered a significant factor in river safety. At low flows, major rapids become difficult to navigate.

Kearsley and Warren (1992) analyzed mooring conditions for white-water boaters at 129 campsites. Mooring conditions were influenced by large fluctuating flows at all sites. This study indicated that better mooring quality exists under constant flows than under fluctuating flows.

Recreation at Lakes Powell and Mead

Lake Powell is the second-largest reservoir in the western United States. Glen Canyon Dam and its power plant were designed to operate between the water levels of 3,490 and 3,700 feet above sea level. In this range the lake has a water surface area of 52,000 to 163,000 acres and a shoreline that is 990 to 1,960 miles long. Lake Powell provides several major categories of recreation: lakeshore and back-country camping, campground use, fishing, boating, beach use, and picnicking. Fluctuations in water level are unavoidable for Lake Powell (Chapter 4). The highest water levels generally occur between April and June and the lowest levels between February and March.

Lake Powell has five marinas, and some expansions and additions are being planned. Normal lake fluctuations influence recreational boating because changing water levels affect access to the water via developed facilities. A change in reservoir levels requires adjustments in facilities, including marinas, docks, buoys and buoy lines, breakwater barriers, channel markers, and ramps.

Boaters use the stretch of Lake Mead where the Colorado River enters the lake for scenic boating, fishing, water skiing, and other recreational pursuits.

Navigability in this interface between Lake Mead and the river may be affected by dam operations.

Recreation and Native American Tribes

A substantial portion of the Hualapai Tribe's gross revenue is derived from river-based recreational activities. The largest of these activities is white-water boating. The Hualapai Tribe owns and operates Hualapai River Runners, a commercial white-water boating company. Hualapai River Runners is one of four enterprises operated by the Hualapai Tribe and was the major source of tribal income in the 1980s. In addition to offering white-water boating trips, Hualapai River Runners provides shuttle services, tows across Lake Mead, and access for river takeouts at Diamond Creek. In 1987 it earned 49 percent of the Hualapai Tribe's gross income.

The tribe has diversified its business interests and now depends less on river-based recreational activities than it did in the past. Nevertheless, the tribe earned about 33 percent of its total 1991 income from such activities. The net economic value of commercial white-water trips that launch at Diamond Creek by arrangement with the Hualapai Tribe was estimated by data from Bishop et al. (1987) on commercial white-water boating in the Grand Canyon. No separate economic value study was conducted for commercial trips launched by the Hualapai Tribe at Diamond Creek.

The Navajo Reservation borders portions of Glen Canyon National Recreation Area and Grand Canyon National Park. There has been little development of business enterprises in this region because of a federal statutory freeze that has precluded construction or development on this portion of the reservation, pending resolution of a territorial dispute. The development ban was lifted recently, and river-based enterprises may develop in the near future. At the present time, however, no river-based enterprises owned or operated by the Navajo Nation have been documented.

Although several other tribes have land bordering Grand Canyon National Park or have current and historical ties to the Grand Canyon, no river-based enterprises owned or operated by these tribes have been documented.

ECONOMIC EFFECTS OF DAM OPERATION ON RECREATION

The effects on recreational activities of various operational schemes for the Glen Canyon Dam are summarized in the operations EIS (BOR, 1994). The EIS gives numerical values where possible; otherwise, it gives qualitative assessments. Assessments are based on rankings of alternative operational scenarios in a study of visitor preferences by Bishop et al. (1987). Each alternative was ranked as more or less favorable for recreation overall and for each of several indicator activities. Indicator activities in the EIS include fishing, day rafting, white-water boating, and lake facilities and activities. Effects of habitat maintenance flows (Chapter 5) are discussed in the EIS under the three alternatives that include such flows.

Background on Economic Methods Used

Two economic measures—the net economic value of recreation and the regional economic impact of recreation—were used in GCES to estimate the national and regional economic effects of proposed alternatives for dam operations.

Table 7.1 summarizes the net economic value of various recreational activities under different types of water release years. The *net economic value* of an activity is its net addition to the nation's output of goods and services, measured in dollars. The net economic value is a measure of the value of an activity above the actual costs of participating in the activity.

Bishop et al. (1987) used the contingent valuation technique to analyze the economic effects of dam operations on recreation. They presented recreationists with descriptions of recreational opportunities at several different flow levels and asked them to state their willingness to pay for these recreational opportunities at different flow levels. Bishop et al. found that the value of angling and white-water boating was related to flow and that there were significant differences between the effects of flow on commercial and private white-water boaters.

Statistical models for angling and commercial and private white-water boating were developed by Bishop et al. (1987) and HBRS (1993). These statistical models describe the relationships among the economic benefits of each recreational activity, the average flow during the month, and the occurrence of fluctuations exceeding 10,000 cfs during the month. For each type of activity, the model calculates net economic benefits per trip and then aggregates benefits over the actual distribution of trips recorded in 1991.

The statistical models predict the same economic benefits for several of the alternatives in the EIS because some alternatives have the same average

TABLE 7.1 Net Economic Value of Recreation (annual benefits in 1991 nominal $ millions)

Type of Release (year)	Anglers	Commercial White-Water Boating	Commercial White-Water Boating Below Diamond Creek	Private White-Water Boating	Total
Low (1989)	1.3	5.4	0.104	1.1	7.904
Moderate (1987)	1.2	6.4	0.122	1.2	8.922
High (1984)	1.1	12.4	0.230	2.0	15.730

SOURCE: BOR (1995).

monthly flows and the same degree of fluctuations over 10,000 cfs. For example, both the Interim Low Fluctuating Flow and Existing Monthly Volume Steady Flow Alternatives have the same average monthly flows. There would be no fluctuations under the Existing Monthly Volume Steady Flow Alternative and no fluctuations over 10,000 cfs under the Interim Low Fluctuating Flow Alternative. Consequently, the statistical models do not distinguish between these two alternatives. Likewise, the No Action, Maximum Power Plant Capacity, and High Fluctuating Flow Alternatives all allow daily fluctuations exceeding 10,000 cfs and have identical average releases.

Much of the white-water boating use occurs during the summer months. Most of the angling use occurs during the spring and fall. These patterns of use have important effects on the generation of net economic benefits. To the extent that net economic benefits are directly determined by flow, changes in flow during periods of high recreational use produce larger changes in net economic value than similar changes in flow occurring at other times of the year.

Regional economic impact is a measure of the importance to the local economy of the expenditures. Since expenditures made by recreationists reflect the costs of participation, they are not considered benefits from the national point of view and are not included in the calculation of net economic value.

River-based recreationists, such as anglers and white-water boaters, spend large sums of money in the Grand Canyon region. Such expenditures provide some measure of the local impacts of recreational users. Direct expenditures alone, however, do not fully measure the effects of spending by visitors to the region. Local businesses and residents spend part of the money they receive from anglers and white-water boaters to purchase goods and services from other individuals and local businesses. These individuals and businesses, in turn, spend a portion of their revenue in the region, and so on. Because a portion of each dollar spent by nonresident recreationists is respent over and over in the region, the effect of each dollar of direct expenditure by visitors is greater than $1. A multiplier relates the amount of nonresidential expenditure to the total amount of local economic activity produced by the visitor's spending. Multipliers allow the effect of nonresident expenditures to be more fully assessed. The U.S. Forest Service's Impact Analysis for Planning model (Taylor et al., 1992), a sophisticated framework for assessing regional effects of expenditures, was used to estimate multipliers for this analysis.

Estimates of average expenditures by anglers and white-water boaters were obtained by Bishop et al. (1987). Commercial white-water boaters generate most of the economic activity in the region. In total, river-based recreational users generated some $23 million in local economic activity in 1991. Because the number of white-water boating trips is not expected to change and the number of angling trips taken is held constant for this analysis, there is no change in regional economic activity for any of the alternatives listed in the EIS.

Economic Effects of Dam Operations on Anglers

The quality of a fishing trip for most anglers in the Glen Canyon reach is highest during moderate steady discharges because such discharges appear to improve several attributes of fishing trips. Presently, there are no constraints on the number of anglers permitted to fish in Glen Canyon. The number of fishing trips to the area in any given year varies with general economic conditions, fishing regulations, and the quality of the fishery.

Anglers using the Glen Canyon trout fishery place a high value on large fish. Under the EIS fluctuating flow alternatives, including no action, trout were assumed to be less likely to reproduce and survive until they reach trophy size. Under the Moderate, Modified Low, and Interim Low Fluctuating Flow Alternatives, the potential for catching large fish was assumed to increase, and therefore fishing trip quality also would have the potential to increase. The underlying validity of this assumption is questionable (Chapter 6), but the economic analysis does show the sensitivity of the value of the fishing to fish population size and age structure.

Rapid changes of stage put wading anglers in Glen Canyon at risk of inundation. This risk would be reduced under the alternatives with ramp rate restrictions and would be eliminated in the steady flow alternatives. Downstream in the Grand Canyon, angler safety is not believed to be significantly affected by dam operations, primarily because most fishing activities take place from boats or from shore.

Studies in other basins have documented a relationship between angling quality and the number of trips taken. In these studies, angling quality has been related to the species, number, and size of fish caught as well as by the presence of native fish in the catch. Some ways of operating Glen Canyon Dam may change these factors for anglers who fish below the dam. A change in the quality of the fishery might result in the total amount of fishing that takes place. Biological models that could predict angling quality are unavailable, however, and economic models that could predict the amount of fishing based on the quality of the fishing have not been developed. As a result, it is not possible to predict changes in the economic value of angling as a function of dam operations.

Economic Effects of Dam Operations on Day Rafting

Boaters in the Glen Canyon reach, most of whom are anglers, have difficulty navigating around sand bars when discharge is 3,000 cfs or less. Most boaters are unable to move up or downstream, and some of those attempting to do so hit rocks and sustain boat and motor damage.

Minimum flows of 5,000 cfs eliminate navigation and safety considerations for most day rafters and other boaters. Steady flows make sand bars passable to all boaters.

All EIS alternatives were treated as having similar influences on day rafting. Also, habitat maintenance flows (occasional high flows) are unlikely to have any effect on the quality of day rafting below Glen Canyon Dam. Because the alternatives do not differ significantly for day rafters, the economic effects of changing operations are estimated as zero.

Economic Effects of Dam Operations on White-Water Boating

White-water boaters prefer moderate fluctuations and steady flows because of their influence on itinerary, character of rapids, wilderness values, and boat management at camp. White-water boaters were asked to rank several operational scenarios in the study by Bishop et al. (1987). Of the EIS alternatives, the steady flow would be most preferred by this group. Fluctuating flow alternatives with daily range and ramp restrictions and 5,000-cfs minimum flows are more acceptable than those without such restrictions.

Wilderness values are influenced by daily fluctuating flows. When the river undergoes wide daily fluctuations, most river runners are aware of these fluctuations and think they make the setting seem less natural (Bishop et al., 1987). Fewer river runners would be aware of more moderate daily fluctuations.

An index of white-water accident risk (Brown and Hahn, 1987) was used to compare the safety of alternatives. At low flows, accident potential is greatest for commercial motor and small oar-powered craft. Risk is reduced most by steady flow; restricted fluctuating flow reduces risk half as much as steady flow.

The accessibility of the river to the handicapped was raised as an issue for the EIS and is a concern for NPS, which issues permits preferentially for trips with handicapped individuals. Effects on accessibility follow the same pattern as general accident risk.

The number, size, and character of camping beaches in the Grand Canyon have a direct effect on the total recreational capacity of the river corridor and the experience for white-water recreationists. Under the fluctuating flow alternatives, the distribution of sites within power plant capacity would be 0.7 sites per mile in narrow reaches and 1.1 sites per mile in wide reaches. Steady flow alternatives would support 0.9 sites per mile in narrow reaches and 1.1 in wide reaches.

The size of a particular camping beach is highly variable in relation to flow. In most years the area suitable for camping would average 7,720 ft^2 or less under the fluctuating flows and up to 9,200 ft^2 under steady flows. Fluctuating flows influence mooring and cause boat management problems and stranding. Under the fluctuating flow alternatives, mooring would be fair to good at 64 percent of camping beaches; under steady flow, 92 percent would be fair to good.

Economic Impacts of Dam Operations on Native American Tribes

A number of commercial and private white-water boating trips launch from Diamond Creek on the Hualapai Reservation. White-water boating use below Diamond Creek, as measured by the number of trips taken, is expected to increase over time until use reaches capacity limits. The nature and timing of this increase cannot be reliably predicted. Changes in the number of trips are expected to be unrelated to dam operations.

SUMMARY OF EFFECTS OF DAM OPERATIONS ON RECREATION

The majority of recreational benefits are derived from commercial white-water rafting, which in general is positively related to average daily flows and negatively related to fluctuations. Alternatives that increase average summer flows or eliminate daily fluctuations in excess of 10,000 cfs tend to increase recreational benefits. The estimates of net economic benefit are based on the statistical relationship between flow and recreation holding all other factors constant at the time of the study. Therefore, these benefit estimates do not account for any long-term changes in the recreational environment that might affect value. Table 7.2 shows expected changes in the equivalent annual value of recreation under different alternatives, as estimated in the EIS (BOR, 1995).

STUDIES OF NONUSE VALUE

The final EIS (March 1995) includes a brief but useful discussion of the conceptual basis for nonuse values, their potential magnitude, the reasons why people hold nonuse values, the resources to which nonuse values may apply, and methods used by GCES to assess nonuse values. Preliminary studies indicated that quantification of nonuse values associated with Glen Canyon Dam operations is feasible. Participants were able to distinguish effects on the river corridor from effects on the Grand Canyon in general and indicated that they, as nonusers, would be affected by changes in dam operations. A full-scale nonuse value study was completed in mid-spring 1995. Findings from the survey of 2,550 households in the Colorado River Storage Project power marketing area and 3,450 households in the United States as a whole have been made available as a GCES report (Welsh et al., 1995). The results, summarized below, show that substantial nonuse values are at stake with regard to managing dam operations.

The nonuse value work examined public values associated with only three main alternatives for operating the Glen Canyon Dam, even though the EIS

TABLE 7.2 Change in Equivalent Annual Value of Recreation for the 50-year Planning Period, as estimated by BOR for the operations EIS[a]

Alternative	Changes in Equivalent Annual Value Compared to No Action (1991 $ millions)				
	Anglers	Commercial White-Water Boating	Private White-Water Boating	White-Water Boating Below Diamond Creek	Total
No Action	0.0	0.0	0.0	0.0	0.0
Maximum power plant capacity	0.0	0.0	0.0	0.0	0.0
High fluctuating flow	0.0	0.0	0.0	0.0	0.0
Moderate fluctuating flow	0.40	0.10	-0.10	0.0	0.40
Modified low fluctuating flow	0.90	2.60	0.20	0.04	3.74
Interim low fluctuating flow	1.00	2.70	0.20	0.04	3.94
Existing monthly volume steady flow	1.00	2.70	0.20	0.04	3.94
Seasonally adjusted steady flow	0.80	3.60	0.30	0.06	4.76
Year-round steady flow	1.00	1.70	0.20	0.03	2.93

[a]The net economic benefits in each year were inflated by the projected gross national product price deflator for that year (Power Resources Committee, 1993) and were discounted using the federal discount rate of 8.5 percent). The equivalent annual value was then calculated using the same rate. SOURCE: BOR (1995).

considered nine alternatives. The three that were analyzed included moderate fluctuating flows, low fluctuating flows, and seasonally adjusted steady flows. These three alternatives were distinct enough to allow respondents to distinguish among them and cover the range of operations within which the preferred alternative falls. Seven different versions of a mail questionnaire were developed and tested. Nonuse values were measured for two samples: a random sample of households located in the area of the Southwest served by Salt Lake City Area Integrated Projects, called the marketing area sample, and a random sample of households nationally. The response rate was 66 percent for the national sample. Nonrespondents were contacted by telephone in an attempt to compare respondents to nonrespondents. The two groups differ from each other in a number of respects and in a manner suggesting that the values held by nonrespondents would be lower, on average, than those held by respondents. Data for nonrespondents were taken from phone surveys, and the number of nonrespondents who would support various proposals to modify dam operations was estimated and incorporated into economic value calculations for each sample group (Welsh et al., 1995).

The report by Welsh et al. exhaustively evaluated the GCES nonuse value studies against criteria developed by economists to examine the validity of nonuse value studies. The GCES nonuse value study performed well when evaluated by standard validity criteria (Welsh et al., 1995).

The nonuse values, as documented by Welsh et al. (1995), are significant when compared to the recreational values and the foregone power revenues associated with modified dam operations (Table 7.3).

A question arises regarding whether the nonuse values from the national sample or the marketing area sample deserves more weight in decisions about dam operations. There is no consensus on this matter. On the one hand, the resources affected by the dam are located in a national park and national recreational area. Thus, the national nonuse values are highly relevant. The results for the marketing area reflect the preferences of those who live close to the study area and thus are more likely to visit the site and who also may incur higher electricity costs as a result of modified dam operations. Consequently, some would argue that the preferences indicated in the marketing area sample should take precedence over the national sample. For either group, however, the benefits of a moderate or low fluctuating flow alternative appear to outweigh the foregone power revenues.

For the seasonally adjusted steady flows, the foregone power revenues are somewhat larger than the combined recreational and nonuse values in

TABLE 7.3 Annual Values Associated with Alternative Dam Operations ($ millions)

			Nonuse Values	
	Power	Recreation	National	Marketing Area
Moderate fluctuating flow	36.7 to 54.0	+0.4	+2,286.4	+62.2
Low fluctuating flow	15.1 to 44.2	+3.7	+3,375.2	+60.5
Seasonally adjusted steady flow	-88.3 to -123.5	+4.8	+3,442.2	+81.4

SOURCE: Adapted from Tables 7-1, 7-2, and 7-3 in Welsh et al., 1995; and as corrected in Table 7-1 in Welsh, 1995.

marketing area. The national nonuse values, however, are about 30 times larger than the foregone power revenues for seasonally adjusted steady flows.

SUMMARY

Studies of recreation economics were designed and conducted using state-of-the-art economic methodologies that are appropriate for the task of measuring the economic impacts of EIS alternatives on recreationists. The CVM was applied in a manner that maximizes the reliability of the recreational value results. Surveys were extensively tested prior to being administered, sample sizes were adequate, and statistical results were robust and consistent with economic theory (Chestnut et al., 1991).

It is important to keep in mind several issues when interpreting the economic analysis of recreation. The analyses focused on the relationship between recreational benefits and the immediate effect of river flows on the quality of recreational experiences. For both the white-water rafters and anglers, other long-term factors are related to the various alternatives and to the quality of the recreational experience. For anglers the implications of alternatives are very uncertain over the long term.

Factors such as the availability of camping beaches play a role in the quality, and thus the net benefits, of rafting trips. The economic analyses, however, focused on benefits associated with trips in which the number and sizes of beaches were fixed, and so the recreational benefits underestimate the long-term benefits associated with alternatives that would maintain larger numbers or sizes of beaches (Chapter 5).

The GCES nonuse value studies are one of the most comprehensive efforts to date to measure nonuse values and apply the results to policy decisions. The studies were subject to extensive scrutiny by the interests (agencies, advocacy groups) participating in GCES and also to intensive review by a panel of professional economists with no stake in the outcome of the studies. While the CVM was applied in a manner consistent with current professional practice for measuring nonuse values, there is no objective standard of benefits against which the CVM results can be compared. If there were, the CVM exercise would not have been necessary. While not completed in time to be reported in the final EIS, the nonuse value results are an important contribution of GCES and deserve full attention as decisions are made regarding dam operations.

REFERENCES

Bishop, R. C., et al. 1987. Glen Canyon Dam Releases and Downstream Recreation. GCES Technical Report, Bureau of Reclamation, Salt Lake City.

Brown, C.A. and M.G. Hahn. 1987. Effect of flows in the Colorado River on reported and observed boating accidents in Grand Canyon, Glen Canyon Environmental Studies Technical Report. Bureau of Reclamation, Salt Lake City, Utah.

Bureau of Reclamation. 1994. Operation of Glen Canyon Dam. Draft Environmental Impact Statement, U.S. Department of the Interior, Washington, D.C.

Bureau of Reclamation. 1995. Operation of Glen Canyon Dam. Final Environmental Impact Statement, March, U.S. Department of the Interior, Washington, D.C.

Chestnut, L., R. Raucher, and R. Rowe. 1991. A Review of the Economic Studies Conducted in Phase I of the Glen Canyon Environmental Studies. Prepared for the Glen Canyon Environmental Studies by RCG/Hagler, Bailly, Inc.

Harpman, D.A., M.P. Welsh, and R.C. Bishop. Nonuse Economic Value: Emerging Policy Analysis Tool." Rivers 4 No. 4 (March 1995):280-291.

HBRS, Inc. 1991. Assessing the Potential for a Total Valuation Study of Colorado River Resources. Final Report, prepared for the Glen Canyon Environmental Studies by HBRS, Inc., Madison, Wisc.

HBRS, Inc. 1993. Analysis of the Impact of GCDEIS Alternatives on Recreational Benefits Downstream from Glen Canyon Draft Report. Prepared for the Glen Canyon Environmental Studies by HBRS, Inc., Madison, Wisc.

Kearsley, L.H., and K. Warren. 1992. (1993 in EIS) River Campsites in Grand Canyon National Park: Inventories and Effects of Discharge on Campsite Size and Availability. Final report. Grand Canyon National Park Division of Resource Management, National Park Service.

National Oceanic and Atmospheric Administration (NOAA). 1993. Report of NOAA Panel on Contingent Valuation. U.S. Department of Commerce, Washington, D.C.

National Research Council. 1987. River and Dam Management: A Review of the Bureau of Reclamation's Glen Canyon Environmental Studies. Washington, D.C.: National Academy Press.

Power Resource Committee. 1993. Power Systems Impacts of Potential Changes in Glen Canyon Power Plant Operations. Glen Canyon Environmental Studies Technical Report, Stone and Webster Management Consultants, Inc., Englewood, CO.

Taylor, C., S. Winter, G. Alward, and E. Siverts. 1992. Micro IMPLAN User's Guide. U.S. Department of Agriculture, Forest Service, Land Management Planning Systems Group, Fort Collins, Colorado.

U.S. Department of the Interior. 1991. Notice of proposed rulemaking: Natural resource damage assessment. Federal Register 56(82):19752-19773.

Welsh, M. 1995. Memorandum on corrections to the GCES Non-use Values Study Draft Final Report, July 28. Prepared by Hagler Bailly Consulting, Madison, Wisc.

Welsh, M.P., R.C. Bishop, M.L. Phillips, R.M. Baumgartner. 1995. GCES Nonuse Value Study. Draft final report, prepared by RCG/Hagler Bailly, Inc., Madison, Wisc. July 12.

8

Cultural Resources

Cultural resources along the Colorado River between Glen Canyon Dam and Lake Mead include physical remnants of human presence prior to the arrival of Europeans and during the era of exploration by Europeans. In addition, cultural resources include sites that have significance to one or more cultures presently, in the recent past, or in the distant past but without necessarily showing any physical evidence of human presence. Finally, cultural resources can include general landscape such as the river, the canyon, or particular kinds of geomorphic or biotic features along the river, especially if they have significant traditional cultural properties. In fact, the entire region has acknowledged cultural significance to both Native Americans and Americans generally.

Cultural resources seem to present an incredibly broad arena for study, even in a specific environment such as the Glen Canyon Environmental Studies (GCES) study area. The task of GCES, however, was not to study all aspects of cultural resources but rather to focus on those particular resources or locations that might be affected by various alternative means of operating Glen Canyon Dam. The distinction between studies of cultural resources generally and studies of resources potentially affected by operations was never clearly established or maintained by GCES. As will become evident in this chapter, failure of GCES to focus its resources on questions related to operations produced lack of specificity in conclusions about operations. At the same time, the tendency of GCES to direct money at virtually any aspect of cultural resources led to inevitable inadequacy in funding of the most relevant cultural resource issues, such as sacred sites.

The GCES organizers took a dual approach to the study of cultural resources. The first studies to be identified and supported through GCES were archaeological, that is, they involved inventory of sites along the river showing evidence of past human presence. At first the work did not involve participation by the Native American tribes having cultural affinities with the Grand Canyon area; the studies were conducted primarily by the National Park Service (NPS) pursuant to various federal laws. A second thrust developed later around ethnography in 1992 when the six Native American tribes or tribal groupings were first acknowledged as cooperators in the guidance of GCES. Ethnographic studies, performed primarily by tribal people or consultants hired through the tribes, dealt with present and historical cultural uses of the lands along and above the river between Glen Canyon Dam and Lake Mead. Both of these categories of study will be discussed in this chapter.

NATIVE AMERICANS IN THE GRAND CANYON REGION

Overview of Present Residents

The populations most directly affected by dam operations are those residing in northeast Arizona, specifically the peoples of Coconino, Apache, and Navajo counties. In 1990 the U.S. Bureau of the Census estimated a total population of slightly more than 235,000 for these three counties. Of this total, Native Americans accounted for 116,463, or 49 percent (Table 8.1).

The tribes of the Grand Canyon region differ greatly in population size. Tribal population figures, however, are available only for individuals residing on reservations. Large numbers of Navajos, Hopis, and Zunis as well as members of the other three tribes live beyond reservation boundaries or in cities and towns well outside the region. The number of individuals who live on the reservations does show, however, approximate relative differences in population size. About 140,000 members of the six tribes live on reservations in either Arizona, New Mexico, or Utah (Table 8.2). Even without including Navajos residing on those portions of the Navajo reservation in New Mexico and Utah, they would still number 87,577 or 85 percent of the total potentially affected Indian population. Collectively, the members of the tribes shown in Table 8.2 constitute the largest concentration of culturally traditional Native American peoples in the United States.

Although there are differences between tribes and individuals, some generalizations can be made. The majority of adult members of these tribes speak their tribal language as their first or, in many cases, only language. Al-

TABLE 8.1 Population Sizes by County

County	White (non-hispanic)	Indian	Total
Coconino	57,170	28,233	99,591
Apache	11,468	47,803	61,591
Navajo	31,148	40,417	77,658
Total	99,786	116,453	235,840

SOURCE: Bureau of the Census (1993a).

TABLE 8.2 Summary of Population Sizes by Tribe

Tribe	Reservation Population	Percent Under Age 18
Navajo (total)	123,944	43.6
Arizona only	87,577	43.9
Hopi	7,061	38.1
Hualapai	801	42.5
Havasupai	400	41.8
Southern Paiute		
Kaibab	102	41.2
Shivwits[a]	85	
San Juan[b]	150	Not available
Zuni	7,073	38.4

[a]The number of "Paiutes of Utah" enumerated in Washington County, Utah. [b]Estimated; the San Juan Paiutes are counted with the Navajos. SOURCE: Bureau of the Census (1993b).

though Christianity and Western medicine have been accepted to some extent among Native Americans, the vast majority, to varying degrees, still adhere to tribal religious and healing practices. Although their numbers are declining, a significant portion, particularly elderly members of these tribes, are still engaged in and dependent in part on traditional economic activities such as herding, farming, gathering, and craft production. Traditional tribal beliefs and practices are still part of the living culture of these peoples. Actions that threaten their beliefs and practices endanger the continued well-being of the communities. The concerns expressed by the tribes relative to the potential effects of dam operations must be considered and evaluated within the context of cultural traditions and values that at times differ significantly from those of most Americans.

The vast Indian landholdings in the region create a false picture of Indian economic well-being and potential. As in most regions of the United States, Indian reservations of the West were established in areas with limited resources, in which early white settlers showed little interest. While reservation lands may appear to be only sparsely occupied to the casual observer, most reservations are overpopulated given their present economic base. In addition, the tribal populations are increasing rapidly. In 1990 the percentage of reservation populations under age 18 ranged from 38 percent to almost 44 percent (Table 8.2).

In terms of income there are significant differences among tribes; the most significant difference is between reservation Indian populations and local white (non-Hispanic) peoples. The Kaibab Paiutes have by far the highest income but number only 102 individuals (Table 8.3). In contrast, the Navajos, who constitute about 85 percent of the total regional Indian population, have a per capita income of only $3,805 and a median family income of only $11,532.

If not for various government programs, these incomes would be even lower. The income of reservation Indians averages less than one-half that of their white neighbors (Table 8.4).

In terms of cultural and historic traditions and religious beliefs and practices, the Native American peoples are the population at risk relative to dam operations. It is also important to note that the relative importance of cultural and religious resources in the canyon varies significantly from tribe to tribe. Also, in terms of the potential economic effects of dam operations, Native American peoples are the poorest and thus the group most at risk within the region. The degree and nature of potential economic effects vary. For some tribes, dam operations have no potential positive or negative economic effects. Others may be affected in important ways.

TABLE 8.3 Indian Income by Reservation (1989)

Reservation	Per capita	Median Family
Navajo (all)	$3,805	$11,532
Hopi	$4,566	$13,917
Hualapai	$3,630	$11,731
Havasupai	$4,112	$20,179
Southern Paiute		
Kaibab	$5,245	$21,250
Shivwits		
San Juan	Not available	Not available
Zuni	$3,904	$15,502

SOURCE: Bureau of the Census (1993b).

TABLE 8.4 Income of White (non-Hispanic) Populations in Three Counties of Arizona

County	Per Capita	Median Family
Coconino	$13,919	$37,761
Apache	$11,694	$34,734
Navajo	$11,731	$31,106

SOURCE: Bureau of the Census (1993a).

Historical Perspective

Archaeological studies have shown that human occupation of the Grand Canyon began as early as 2000 B.C. About A.D. 700 horticultural Puebloan peoples began to settle in the canyon. Cultural characteristics as well as recorded traditions support the interpretation that at least some of these Puebloan peoples were ancestral to the modern Hopis. About A.D. 1200 the Puebloan settlements in the canyon were abandoned. Most of these Puebloan peoples eventually settled on the Hopi mesas (Heib, 1979; Clemmer, 1995). Based on archaeological evidence, anthropologists have long noted what appears to be a cultural historical affiliation between Zuni and the prehistoric Puebloan peoples of the Chaco Canyon region of New Mexico (Woodbury, 1979). Only recently have the Zuni begun to assert that they originated in the Grand Canyon area and migrated up the Little Colorado River to eventually settle along the Zuni River (Ferguson and Hart, 1985). Although they are no longer residents of the Grand Canyon, the Hopis used sacred sites in the canyon for religious purposes and continue to do so today (Clemmer, 1995). In recent years the Zuni also have asserted continued use of sacred sites in the canyon (Ferguson and Hart, 1985).

After A.D. 1300 small groups of non-Puebloan peoples began to occupy the Grand Canyon at least seasonally. The occupants of the entire north side of the canyon and the south side as far west as the Little Colorado River appear to have been ancestors of the modern Southern Paiutes. On the south side of the canyon as far east as the Little Colorado, the new occupants were the ancestors of modern Hualapais and Havasupais. During the nineteenth century, the Southern Paiutes were forced by Euro-American settlers to abandon their croplands in the canyon. The Shivwits and Kaibab were placed on reservations some distance away. These groups claim, however, continuing use of specific sites in the canyon for religious purposes (Stoffle et al., 1993). Sites claimed by Southern Paiute in Stoffle et al. (1995) are, according to Grand Canyon National Park archaeological site records, Anasazi A.D. pre-1150 and Pai (Hualapai and Havasupai) A.D. post-1300. The Hualapais also continue to assert religious use of the canyon (Hualapai Cultural Resources Division, 1993).

The Navajos were the last Native American tribe to enter the region. The significance of their history of occupancy is the most uncertain. During most of the historical period, the Navajos were primarily a pastoral people. Divided into numerous small, highly autonomous extended families and clans, Navajo family groups wandered widely in search of water and forage for their herds. During the eighteenth and nineteenth centuries, the Navajos were the only tribe in the Southwest to experience a major population increase. From only a few thousand in the mid-1700s, they grew to over 10,000 by the mid-1800s and to over 20,000 by 1900 (Johnston, 1966). Population growth, together with their ability in times of drought and war to temporarily move great distances in search of water and safety, resulted in their expansion and fluid territorial boundaries. Territorial expansion was also the major source of their conflict with neighbors, particularly the Hopi and Zuni. During the Navajo War of 1863-1868, a number of Navajo families sought refuge in the Grand Canyon. Some scholars argue that Navajo families were already occupying at least portions of the region as early as the 1700s (Thomas, 1993), while others believe that it was not until after 1880 that the Navajos permanently settled in the region (Bunte and Franklin, 1987; Euler, 1974).

Adding to the confusion over interpretations of Navajo concerns is the question of tribal definition. The Navajos together with the Apachean tribes of the Southwest are Athapaskan speakers. The Athapaskan-speaking peoples were the last Native Americans to arrive in the region; their arrival may have been as late as A.D. 1500. Thus, many individuals, including some scholars, tend to portray the Navajos as late invaders of the area.

While the Navajos speak an Athapaskan language, the Navajo population growth during the eighteenth and nineteenth centuries was in large part the result of the absorption of numerous non-Athapaskan individuals and families. The Navajo tribal study (Roberts et al., 1994) notes several examples from the Grand Canyon area. Possibly one-third of the Navajo clans and subclans originated from incorporation of non-Navajo peoples, including Hopis, Paiutes, Zunis, and Utes. The Navajos not only assimilated these peoples but also incorporated many of their cultural and religious beliefs and practices into what has been an ever-expanding but still uniquely Navajo cultural and religious tradition. Thus, biologically and culturally, the contemporary Navajos are a fusion of Athapaskan and earlier southern peoples (Bailey and Bailey, 1986; Vogt, 1961; Reichard, 1928).

Background Information on the Six Tribes

Hualapai

Historically, the Hualapais were primarily a hunting and gathering people who occupied much of northwestern Arizona south of the Grand Canyon. While some Hualapai families farmed small plots in the side canyons of the Grand Canyon, most lived in widely scattered seasonal camps. In 1883 the present Hualapai reservation was established by presidential executive order (Dobyns and Euler, 1974; Kappler, 1904) and included in the reservation the south bank of the Colorado River for a distance of some 108 miles. Traditionally, the Hualapais have claimed territory extending to the middle of the Colorado River. The 1990 census recorded 802 Native Americans resident on the Hualapai reservation (Bureau of the Census, 1993b). A tribal enterprise, the River Runners, operates raft trips on the Colorado for tourists. While the Hualapais are diversifying their tribal economic development, as late as 1991 the tribe earned approximately one-third of its total income from river-based recreational activities.

Havasupais

Culturally and linguistically, the Havasupais are a band of the Hualapais. In fact, the Hualapais count the Havasupais as one of their 14 bands. Today, however, they exist as a separate tribal entity recognized by the federal government. The major factor that historically has distinguished the Havasupais from the Hualapais is their occupation of the small but relatively rich farmlands in Cataract Canyon, a side canyon of the Grand Canyon. As a result, the Havasupais have had a much greater dependence on farming than their Hualapai kin. A reservation for the Havasupais was established by executive order in Cataract Canyon in 1880 and 1882 (Kappler, 1904), and in 1975 Congress enlarged the reservation to encompass areas on the adjacent plateau. Today, about 400 tribal members are residents of the reservation (Bureau of the Census, 1993b). While tourism generates some income for the tribe, the reservation boundaries do not extend to the river, and tourism is not directly related to recreational use of the river.

Southern Paiute Consortium

The consortium originally included four distinct groups of Paiutes: the Shivwit Paiutes, the Band of the Paiute Tribes of Utah, the Kaibab Paiutes, and the San Juan Paiutes. However, in 1994 the San Juan Paiutes withdrew from participation in the GCES cultural studies due to other tribal business. They requested the right to reenter cultural resource discussions in the future. Each of these three groups is recognized by the federal government as a separate tribal entity. Closely related culturally, socially, and linguistically, these three communities were part of the Southern Paiute tribes. Historically, the Southern Paiutes had exclusive use of the entire north bank of the

Colorado River as well as portions of the south bank between Glen Canyon Dam and the Little Colorado. Until the late 1800s, the Paiutes depended on small farms in the Grand Canyon as well as on hunting and gathering in the adjacent plateau. Today, the Shivwits (about 85 resident members) live near St. George, Utah (Bureau of the Census, 1993b). The Kaibab (about 100 resident members) have a reservation near Fredonia, Arizona (Bureau of the Census, 1993b), and the San Juan Paiutes (about 150 members) reside in two small communities on the western Navajo reservation (see Bunte and Franklin, 1987). The San Juan Paiutes were legally recognized by the federal government as part of the Navajo Tribe until 1989, when they were accorded separate tribal recognition. The federal government has yet to address the issue of a separate land base or reservation for the San Juan Paiutes (U.S. Department of the Interior, 1989). Paiute residence areas or reservations today are 25 to 75 air miles from the river. There is no indication that operation of Glen Canyon Dam will directly affect the economy of these three tribal entities.

Hopis

A farming people, the Hopis live in a series of permanent villages stretching from Moenkopi on the west to First Mesa on the east, a distance of about 70 highway miles. The Hopi reservation was created by executive order in 1884. The reservation was not created exclusively for Hopi use, however. This has resulted in a long unresolved land dispute with the Navajos, whose reservation surrounds the Hopi lands (Kappler, 1904). Today, about 7,000 Hopis reside on their reservation (Bureau of the Census, 1993b). The closest Hopi village to the river is Moenkopi, about 30 air miles east; the intervening land is part of the Navajo reservation. While tourism is still important to the Hopi economy, it is unrelated to recreational use of the river, and dam operations appear to be of little economic concern to the Hopi Tribe.

Navajos

Historically, the Navajos have been a pastoral people who practiced some farming. With the total population of almost 125,000 members on reservation lands and an almost equal number living elsewhere, the Navajo are the largest tribe in the United States (Bureau of the Census, 1993b). They also have the largest reservation in the United States. It encompasses much of northeastern Arizona, northwestern New Mexico, and a small portion of southeastern Utah. The original Navajo reservation, created by the Treaty of 1868, was a relatively small area along the New Mexico-Arizona border. Since 1868, the reservation has been expanded on a number of occasions by executive order. By executive orders in 1884, 1900, and 1930, the Navajo reservation was extended westward to the Colorado River (see Williams, 1970; Kappler, 1904). In 1969, however, the Solicitor's Office of the U.S. Department of Interior ruled that a 1917 executive order withdrew Marble Canyon for "water power purposes" and placed the Navajo reservation boundary one-quarter mile from the river. Although the Navajo tribe disputes the ruling, Marble Canyon is presently administered by the National Park Service. In addition, the Navajos are the only tribe whose reservation adjoins Lake Powell. Virtually the entire south shore of Lake Powell is Navajo reservation land.

As a result of a land dispute between the Navajo and the Hopi over portions of the western extension area, in 1966 Commissioner of Indian Affairs Robert Bennett ordered a freeze on economic development projects on the westernmost portion of the Navajo reservation until the land question was settled. The "Bennett Freeze" remained in effect until 1992 and then was reimposed in 1995 (Clemmer, 1995).

The Navajo Tribe has voiced a range of economic concerns relative to the operation of Glen Canyon Dam. The Navajo Tribal Utility Authority, provides electricity to the majority of consumers on the Navajo reservation. The authority receives about 20 percent of its power from the Western Area Power Administration (WAPA). Navajo Agricultural Products, Inc., a tribal enterprise that operates the Navajo Indian Irrigation Project, which receives its power from WAPA—by 1998 this will amount to 96 MW. Thus, the Navajo Tribe has a direct interest in the cost of electricity from Glen Canyon Dam (Thomas, 1993). The tribe has also voiced concern over tourism and recreation in Glen Canyon and Grand Canyon. These activities benefit local tribal members, many of whom have small businesses along the highways. The Bennett Freeze restricts economic development that would allow the Navajos even greater opportunity to earn tourist dollars. The tribe also sees the opportunity to develop commercial rafting and sport fishing businesses on the river and has plans to develop a marina on Lake Powell (Thomas, 1993).

Zunis

Historically, the Zunis have been a farming people whose permanent villages have been located near the headwaters of the Little Colorado River drainage, in the extreme western portion of New Mexico. With over 7,000 resident tribal members (Bureau of the Census, 1993b), the Zuni reservation is about 250 air miles from the Colorado River. Dam operations appear to have no direct economic consequences for the Zunis.

ARCHAEOLOGICAL STUDIES

The earliest evidence for human use of the Grand Canyon goes back 3,000 to 4,000 years ago. At that time, Indians were hunting in the canyon, as shown by small wooden split-twig animal figurines left in caves high in the cliffs. These Indians may have been related to the Pinto Basin hunters of the Mohave Desert, who existed about the same time. In all probability the figurines are some form of imitative magic: if a figurine was made of the animal to be hunted, perhaps the maker would have more success in the hunt.

There is no further evidence of human use of the Grand Canyon until about A.D. 500 to 700. At that time, two unrelated groups made halting explorations of the canyon. Along the South Rim came the Cohonina, who practiced minimal agriculture along with hunting and gathering. These Indians lived in harmony with their Anasazi (or Hisatsinom) neighbors to the east. The two groups traded with one another; the Cohonina especially valued the decorated ceramics of the Hisatsinom and adapted their architecture as well. (Hisatsinom is the Hopi term for Anasazi.)

The Hisatsinom occupied both north and south rims from ca. A.D. 500 until around A.D. 1150 to 1200. Their lifestyle was similar to but more sophisticated than that of the Cohonina. They lived in well-constructed masonry pueblos, which occasionally included a subterranean circular ceremonial room, or kiva.

By about A.D. 1050, hundreds of Cohonina and Hisatsinom were living in and around Grand Canyon. Of more than 2,000 archaeological sites now recorded, about 1,500 were inhabited in the twelfth century by the latter group.

Within 100 years, the Grand Canyon was abandoned by the Cohonina and Hisatsinom. Climatic changes were probably a major cause but not the only cause. The Hitsatsinom moved east and became the Hopi as we know them today. The Cohonina simply disappeared from the archaeological record.

For a century or more, no human beings lived in the Grand Canyon. Then, about A.D. 1300, the Cerbat peoples, who were direct ancestors of the present-day Hualapai and Havasupai (the Pai), came into the area from the west, and the Southern Paiute moved into the North Rim. These peoples hunted and gathered wild food and engaged in some farming near springs on the rims and in the canyon depths. They remained until forcibly removed to reservations by U.S. government conquest in the late nineteenth century. Not until the 1880s did any Navajos come to Grand Canyon (Jones and Euler, 1979).

The foundation for understanding the importance of and potential vulnerabilities of archaeological resources is given in the report, *The Grand Canyon River Corridor Survey Project: Archaeological Survey Along the Colorado River Between Glen Canyon Dam and Separation Canyon* (Fairley et al., 1994). This survey, which was supported by GCES, includes all sites that could be discovered within approximately 10,506 acres along 255 miles of the Colorado River from the Glen Canyon Dam to Separation Canyon. A total of 475 archaeological sites and 489 isolated occurrences of cultural materials were recorded during the survey. In addition, direct effects from "recent" inundation were recorded at 33 sites and indirect effects in the form of bank retreat, accelerated arroyo cutting, and increased visitation were noted at 127 sites. This detailed report is marred by only a few errors of fact and field technique. For example, it discusses the controversy regarding the ancestry of the historic Indian occupants in Grand Canyon, but this controversy has been settled and is no longer relevant; archaeologists who have done the most intensive research in the Grand Canyon are in general agreement as to the ancestry of the present tribes.

The most important defect in technique for the archaeological survey was its "no-collection" policy. Artifacts were not collected for future laboratory analyses. The report indicated that this was a Bureau of Reclamation (BOR) requirement, but it is contrary to procedures normally followed by NPS archaeologists at Grand Canyon National Park (J. Balsom, personal communication, May 17, 1994).

Because of the no-collection policy, archaeologists had to accomplish identifications and rudimentary analyses of ceramics and stone artifacts (lithics) under difficult conditions in the field, where such artifacts often were encrusted with dirt or lime accretions. The report notes (p. 21) that the field crews did collect "sherd nips," which are small pieces of the sherd generally less than a 1 x 1 cm square. The report claims that these nips have "as much information potential as the whole sherd for lab analyses" and that field analyses conducted by several different archaeologists could be "re-checked in the lab by a single analyst."

Although some government agencies now demand a no-collection policy in field surveys, the limitations of the policy are very severe. In the opinion of many archaeologists, field examination of artifacts falls far short of accepted scientific analytical procedures in archaeology and prohibits future analyses as new data and theoretical constructs may become available. Further, it leaves the artifacts on the ground where, especially in the Grand Canyon, removal by untrained unauthorized individuals may irretrievably reduce and skew the data base.

The impossibility of laboratory analysis from "sherd nips" becomes apparent in another way. Archaeologists on the project had various degrees of expertise. This is clear from their inability in many cases to distinguish between Pai ceramics (Tizon Brown Ware) and those of the Southern Paiute (Southern Paiute Brown Ware). This resulted in the cultural affiliation of some sites to be listed only as "Pai/Paiute." If whole sherds had been brought to the laboratory where they could be examined by one or more archaeologists familiar with artifact types, this could have been eliminated.

In briefly discussing archaeological aspects of contemporary tribes, the authors (Fairley et al., 1994, p. 110) noted that no Navajo sites or diagnostic artifacts predating the late nineteenth century were identified during the survey. This finding is discussed later in this chapter. Strangely, the authors also noted (p. 111) that little ethnographic material is available for the Pai (Hualapai and Havasupai) bands. These bands have been described in detail by Dobyns (1956) and Dobyns and Euler (1970), which are not referenced by Fairley et al.

Despite these minor caveats, the archaeological survey report is well organized and presents important data on sites along the river corridor. Its value is enhanced by two studies relating archaeological sites to surficial geology in the eastern Grand Canyon. The first report (Hereford et al., 1993) presents evidence of archaeological sites (Pueblo I, ca. A.D. 800 to 900) buried under later ruins (Pueblo II, A.D. 1050 to 1150) that are now being eroded. It also notes even earlier eroding Anasazi ruins marked by hearths, the charcoal from which has been radiocarbon dated at 400 B.C. to A.D. 450.

The archaeological report concludes with the statement that of the 475 known archaeological sites in the river corridor, 50 percent (238) are associated with alluvial deposits, and about 52 percent (123) of these are presently affected by arroyo cutting because of tributary-channel adjustment to hydrological conditions following construction of Glen Canyon Dam.

Field observations along the river in the eastern Grand Canyon attest to the channel erosion that has occurred at archaeological sites since the completion of Glen Canyon Dam. If the geological studies are accurate, large-scale mitigation efforts, including excavation by the NPS, are overdue. There is no indication in any of these studies that NPS has put into action plans or funds for such mitigation. Indeed, the Hopi Tribe (Dongoske, 1993b) has criticized NPS for doing nothing to mitigate these deleterious conditions.

The second report on surficial geology (Hereford, 1993) documents by means of a map and description of the map units the nature of postdam erosional effects on archaeological sites in the area of Palisades Creek. There appears to be no question but that erosion and other alluvial perturbations have taken their toll. Assessments through monitoring should determine which sites are of sufficient significance to warrant mitigation.

Monitoring is documented in another report on archaeological sites from Glen Canyon Dam to the Paria Riffle in Glen Canyon (Burchett, 1993). Of a total of 50 sites, some 38 were studied for damage by erosion and human activity (Fairley et al., 1994). This is a very good report that contains data on how the sites are being monitored, the condition of the sites, and what is recommended for mitigation. Such documentation should be provided for sites along the river in the Grand Canyon as well. At the same time, actual mitigation is overdue.

One undocumented and overly generalized statement does appear in the report, however: "The canyon has been and continues to be a spiritual resource for many cultures, as indicated by various shrines and rock-art sites scattered along the river corridor" (p. 4). That some tribes claim a spiritual resource in the canyon and that there may be known shrines associated with this resource, such as the Hopi salt deposits and sipapu, are not disputed. More specific and documented statements are, however, required as a basis for evaluating cultural resources.

ETHNOGRAPHIC STUDIES

Perspectives on Ethnography

Cultural resources as defined by the archaeologists and cultural resources as defined by the tribes are not the same. To varying degrees and in different ways, all of the tribes see the canyon in its entirety as a sacred place and thus as a cultural resource in need of protection. Within the canyon there are specific locales or sites with what might be best termed secular-historical importance to a particular tribe or group within a tribe. In speaking of the locations of an old sheep corral and a sweathouse, the Navajo study states: "What is most important is the preservation of the stories about those places that perpetuates the significance of the place from generation to generation. Even though the sheep corral . . . or the sweathouses . . . no longer exist, the places where they used to be are still important reminders of the events in Navajo history. Preserving the physical places as much as possible and enabling Navajo access to them helps the perpetuation of the stories and thereby helps preserve the importance of the places" (Roberts et al., 1994, pp. 111-112). In societies for which history exists in the mind and not on paper and is transmitted orally from generation to generation, the physical existence of places is critical for the retention and transmission of historical knowledge. Thus, the orally transmitted stories are the true "cultural resource" and not the physical evidence of human occupation and use.

Tribes also view native plants and animals as part of the cultural resources of the canyon. In this regard, "the loss of native plants is viewed by the Southern Paiute representatives as more damaging than the potential loss of archaeological sites in the banks" (Stoffle et al., 1995).

Sacred Sites

Even more difficult is a special category of cultural resources that, for lack of a better term, is called "sacred sites." Within the canyon specific locales have sacred significance. These sacred sites are the areas of greatest concern to local Native American communities.

The American Indian concept of a "sacred site" has little equivalent to western Judeo-Christian religious tradition. These fundamental differences between Native American and western religious traditions were well defined by the Hopi tribe when they wrote:

> Fundamental to Hopi religious thought is the belief in the sacred nature of physical places such as mountain peaks, springs, and burials. In many religions, including Christianity, the locations of

> most places of worship are theologically irrelevant. The loss of a particular church does not diminish the efficacy of the belief system. In contrast, the loss of a sacred site can damage the vitality and coherence of Hopi religion. Deities are thought of as inhabiting specific locations, and specific geographic areas are identified as points of tribal origin. In such places, individuals interact with deities and the spiritual forces embodied in the natural environment. These interactions are structured by rituals that prescribe the use of particular native plants, animals, and minerals. Activities that may affect sacred areas, their accessibility, or the availability of materials used in traditional practices are of concern to the Hopi. (Hopi Tribe, 1990, p. 1)

Hopi beliefs concerning sacred sites are almost identical to those of the Zunis and to lesser degrees can be applied to the Navajos, Hualapais, Havasupais, and Southern Paiutes as well. The protection of sacred sites is critical to the continuation of religious beliefs and practices. Religious leaders, particularly among the Hopis and Zunis, find themselves in an extremely difficult position. Sacred teachings of both tribes speak of the Grand Canyon as the point of origin or the place of tribal emergence into the present world. They indicate that some of the most sacred sites for the tribe are located in or near the canyon. Destruction of these sites could jeopardize the continuity of their traditional religious beliefs and practices. Disclosure of the locations of sacred sites or of sacred knowledge concerning these sites, however, presents a major problem, particular for Hopi and Zuni religious leaders.

Hopi and Zuni religious traditions are not the same as western religions regarding dissemination of sacred knowledge. Sacred knowledge is not public knowledge, even within the tribes. Religious leaders among the Hopis and Zunis are members of formally organized religious groups or priesthoods. Sacred knowledge is restricted to religious leaders or initiated members of a particular clan or tribal priesthood. This characteristic is not unique to the Hopis and Zunis. Control of sacred knowledge by clan or tribal priesthoods was a common characteristic of the traditional native religions of the tribes of the eastern United States as well. Uninitiated members of the tribe are not privy to certain categories of sacred knowledge, nor are individuals outside the tribe. In large part the authority of the religious leaders of these tribes is derived from their exclusive control of this sacred knowledge. Making public such knowledge could erode or weaken the authority of the religious leaders.

Still another problem that is relevant to the disclosure of locations of sacred sites as well as those of other tribal cultural properties is that of security. This concern of the tribes for the security of cultural resources makes them highly reluctant to disclose information on cultural properties. Most of these sites are located in isolated areas, and looting or desecration of such sites has been a major problem in the Southwest. American Indian objects command high prices. Thus, public identification of the locations of sites could make them targets for thieves in search of objects to sell. Even petroglyphs and pictographs are not safe because rock faces can be removed. Knowledge of the sacred importance of such sites only enhances the market value of objects looted from them. While the Hopis in particular have been victimized by art thieves trading in stolen sacred masks and other sacred objects, this problem is common to all tribes in the region. Still another problem is what some have termed "new-age" people, non-Indians, who are attracted to Indian religious beliefs and some of whom practice a pseudo-Indian religion. Such individuals have attempted to participate in religious rituals and have even left their "prayer sticks" and other "religious offerings" at tribal shrines. Finally, zealous tourists in search of an exotic adventure may intrude on and damage sacred sites (Clemmer, 1995).

Thus, traditional tribal leaders experience a dilemma concerning the protection of their cultural resources. On the one hand, such locations do need protecting. Public disclosure of site locations and the cultural significance of sites may, however, expose them to looting and desecration.

The Hualapai Tribe

The report of the Hualapai Tribe to GCES (Hualapai Cultural Resources Division, 1993) is quite thorough. With the assistance of Hualapai tribal members, the study was written by Robert Stevens of the School of Social Sciences at the University of California, Irvine; Professor Stevens is not a Hualapai. The initial sections presenting geographical, historical, and social background information on the tribe contain elementary data with only a few factual errors. For example, the claim is made (p. 5) that the Cohonina, a prehistoric entity, were ancestors of the Hualapai. This has long been refuted by anthropologists intimately familiar with the matter (Euler, 1981; Schwartz, 1989). The ancestors of the Hualapai were people of the Cerbat tradition.

The Hualapai study also lists significant geographic locations (pp. 11-12), some of which are plainly outside the aboriginal Hualapai range. Examples include Jerome and Paulden in Central Arizona (Dobyns and Euler, 1976). The report states that "in Hualapai worldview, the Grand Canyon system is believed to be the place of emergence of the Hualapai bands." Anthropologists who have researched this (Dobyns and Euler, 1961, 1976) understand that the Hualapai believe their ancestors emerged from a place near Eldorado Canyon on the lower Colorado River southwest of Grand Canyon and then moved to Matawidita Canyon, a tributary of the Grand Canyon from the south and a sacred place in Hualapai tradition. There is a sacred cave, excavated in support of the Hualapai land claims case, called *Wa'ha'vo*, where the legendary chief *Wakiasma* is buried. The importance and sacredness of this site are well documented (Euler, 1958).

The remainder of the report contains statements by older Hualapai regarding the Grand Canyon. At this late date, it may be difficult for individuals to recall details about aboriginal life. The report also lists places, plants, mammals, amphibians, and reptiles that are termed "culturally significant." This is a fairly inclusive listing, and the fact that these places, animals, and plants were known to or used by Hualapai does not say much about the Grand Canyon and the Colorado River.

The intensive research carried out among the Hualapai in the 1950s in connection with land claims cases of the Hualapai (Dobyns, 1956; Euler, 1958) did not yield any emphasis on the Grand Canyon nor was it (other than Matawidita Canyon) considered sacred by respondents. That "the Grand Canyon in its entirety is considered a sacred area by the Hualapai cultural scholars," as stated in the report, may not be literally correct. Respondents may have confused sacredness with animistic belief (i.e., that everything contains a spirit being). It seems obvious, however, that the Hualapai Tribe and its investigators put much thought and effort into their research and report. As shown by their own statements (p. 122), they need more information about the operation of Glen Canyon Dam and about the potential effects of dam operations (p. 137).

Southern Paiute

The Paiute originally held more than 600 miles along the Colorado River from east of the Kaiparowits Plateau in Utah to Blythe, California, as verified by anthropologists who have researched the ethnohistory of the several Southern Paiute bands, including the Chemehuevi (Kelly, 1934, 1964; Euler,

1966, 1972).

A lengthy draft report gives the Southern Paiute assessment of the Colorado River corridor (Stoffle et al., 1993). This report covers the research and study designs, ethnography of the constituent Paiute political units, and the tribal concerns for natural and cultural resources. A separate chapter discusses legal relations between the Southern Paiute for the Colorado River corridor, the Havasupai, and the U.S. government. The section on Southern Paiute place names is very complete and well documented, as are data regarding plants, animals, minerals, trails, and river crossings. The chapter detailing the chronology of Southern Paiute is an excellent ethnohistorical summary.

Southern Paiute "elders" who took three river trips in 1992 and 1993 under GCES sponsorship include some from the Kaibab, Shivwits, and San Juan Paiute tribes. These trips yielded information about tribal interests along the river. The archaeologists in the 1991 Colorado River corridor survey (Fairley et al., 1994) delineated 18 Paiute sites and an additional 32 in which they were undecided about cultural affiliations and described them as Pai/Paiute. Not all of these were visited by the Paiute elders, however. The Paiute report (Stoffle et al., 1993) states that some Paiute continue to use sites along the Colorado River and that most sites visited by the elders "were perceived to be of high cultural significance to Southern Paiute people" (p. 33). Validation of this statement would require evidence and documentation.

The report emphasizes the uses of plants by the Paiute (ethnobotany). The Paiutes undoubtedly used most of the plants listed in the report, but many of these plants are present on the uplands above the river corridor and the Paiute do not necessarily have to go into the canyon to obtain them.

The Paiutes who visited sites in the canyon were cautious about making policy statements; they were uncertain about what the water release options meant. Their conclusions included requests to protect archaeological sites in several ways and a plea for better consultation between federal agencies and the Southern Paiute tribes. They also requested better protection of plants in the canyon. Finally, a four-page recommendation for mitigation of Paiute resources clearly stated the Paiute position as elucidated by the authors of the report.

A second report (Stoffle et al., 1995) dealing with rock art lists 25 sites in the Grand Canyon corridor that have been interpreted by tribal members as Southern Paiute. In addition, rock art at 12 sites in Kanab Canyon, away from the river corridor, and six traditional cultural properties were interpreted by the Paiute as having some relation to their ancestry. Since this is a draft report and perhaps subject to further alteration, all that can be said at present is that it contains Southern Paiute impressions that may have little or no relation to anthropological or historical data.

Hopi

The reports emanating from the Hopi Tribe (Dongoske, 1993a, 1993b) indicate their progress in evaluating cultural resources but provide no conclusions as yet. Mentioned in these progress reports are two studies: a draft historical report entitled "Hopi and the History of Grand Canyon Exploration" by Gail Lotenberg and a cultural resources inventory of the lower Little Colorado River from Blue Springs to the confluence with the Colorado. Neither of these had been released as of September 1995.

Navajo

A Navajo Nation position paper (Thomas, 1993) provides a good summary of the history of laws and dam operations. It claims, without presenting any evidence, that "cultural . . . resources of the Navajo Nation . . . including archaeological sites [and] traditional cultural properties . . . are directly affected by dam operations" (p. 3). In this connection it is worthwhile to again note that the intensive archaeological survey of the river corridor by Fairley et al. (1994) did not locate any Navajo sites.

The Navajo position paper does state that research is being conducted "to document historic and current use and traditional cultural properties of the Navajo people in Glen and Grand Canyons" (p. 4) and that the Historic Preservation Department planned to submit a technical report to the BOR in September 1993. This study was not available for review as of September 1995.

The position paper continues with more undocumented claims of Navajo sites in the Grand Canyon. It says that "many of these archaeological sites [upstream from the confluence of the Little Colorado at Mile 61] exhibit use by Navajo peoples . . . [and that] Navajos have left evidence of use as far west as Crystal Creek at river mile 98" (p. 11). Carrying these claims a bit farther, the position paper claims that traditional cultural properties of the Navajo "reflect Navajo use of the canyons over hundreds of years and the importance of the canyons for their spiritual well being" (pp. 11-12). Again, all that can be said at present is that some empirical evidence must be presented in support of the validity of these statements.

Zuni

There were no extensive reports from the Zuni as of September 1995. Letters and brief reports do provide some information on GCES work by the Zuni.

A letter from Roger Anyon to David Wegner dated January 22, 1993, states that "we anticipate that sites of cultural importance to Zuni will be identified along both the mainstem (Colorado River) and the Little Colorado River." Also, "we are confident that many of the cultural resources identified by archaeologists within the area affected by dam operations are culturally affiliated with the Zuni Tribe. . . ." The Zuni have requested that biologists assist them in protecting natural resources that are identified as having significance to the tribe's traditional and religious concerns. Still, no Zuni archaeological sites have ever been documented along the Colorado River or in the Grand Canyon. In certain aspects of Zuni mythology, there is reference to Zuni emergence from the bottom of Grand Canyon, perhaps from the same place referenced by the Hopi, and it may well be that Zuni assimilated this idea from Hopi.

A letter from E. Richard Hart (Institute of the North American West) to Roger Anyon, dated January 23, 1993, proposes an "exhaustive" search of primary and secondary published materials to produce an annotated bibliography of sources related to Zuni and the Colorado River, Little Colorado River, and Grand Canyon. Mr. Hart states that he has collected information from Zunis on the Grand Canyon for 25 years. But he also notes that Zunis do not reveal religious information and thus "much information in historical and anthropological reports published in the past has contained errors of fact, interpretation, and opinion."

In a five-page report, *Zuni and the Grand Canyon*, dated June 24, 1993, E. Richard Hart indicates that he is now completing "an exhaustive search of primary and secondary published materials as well as a complete search of available manuscript materials" relating to the Zuni claim. This has not been presented to the Committee. He also reiterates that much information in print is faulty or incomplete and states that "much of the published documentary material relating to Zuni emergence and migration is suspect." According to Hart, the Grand Canyon is sacred and plays a prominent role in Zuni religion and world philosophy. His report also claims that periodic pilgrimages to the Grand Canyon are or were made by the leaders of certain Zuni religious groups to obtain samples of items needed for their religion and, finally, that the ecosystems of the Grand Canyon "remain integrated in Zuni religion and greatly influence the religious practices of Zunis today." Evidence of these assertions has yet to be presented.

A report entitled *Zuni Cultural Resources and the Grand Canyon* by Roger Anyon and Andrew L. Othole (four pages, dated June 24, 1993) indicates that the Zuni emergence was from the bottom of the Grand Canyon and that there was a subsequent search for the center of the world, the Middle Place. Then, according to this report, the Zunis moved up the Colorado River and along the Little Colorado. The authors then state that, since Zunis emerged in the Grand Canyon, "all culturally affiliated cultural resources in the Grand Canyon are important to Zuni traditional and cultural values because of the spiritual linkage to the place of emergence for the Zuni Tribe." It also claims that over 400 archaeological sites along the river corridor have significance to the Zuni Tribe. On a recent river trip, Zuni religious leaders visited 28 sites and identified two previously unrecorded Zuni shrines, each on a different site. They state that there may be more but that they cannot be recognized by non-Zunis; they can be identified only by religious leaders. In contrast to these assertions, intensive archaeological research in the Grand Canyon has yet to produce any independent verification of a Zuni presence there.

A three-page progress report by the same authors (May 5, 1993, to September 30, 1993) refers to background research on ethnohistory, preparation of an annotated bibliography, fieldwork, and preparation of an ethnohistorical report. It also records a Colorado River trip taken by the "Zuni team" in May 1993 on the basis of which Zuni ancestral sites and Zuni shrines were identified in the Grand Canyon.

Perspective on Tribal Studies

Although GCES began in 1983, it was not until 1990 that the first cultural resource studies were funded. Why the BOR waited so long to begin supporting tribal studies is not clear, given that most of the laws protecting cultural and native religious properties were already in effect by 1983. Even without any studies, it should have been readily apparent to the bureau that in terms of cultural properties the local tribes were the only population at risk and that in terms of potential economic effect some of these tribes were also the population most at risk. The tribes are an integral part of the ecosystem. It is unfortunate that they were brought into the GCES very late.

After the decision was made to include the tribes, the BOR began contracting with them before it had clearly defined the nature of studies appropriate to define tribal concerns. As the Navajo Nation notes, "Tribal issues in the EIS [environmental impact statement] seemed to be considered synonymous with archaeological and cultural resources" (Roberts et al., 1994). As the Navajo researchers discovered on their own, other issues also were of critical concern to the Navajos.

Even within this general category of archaeological and cultural resource studies, the BOR appears to have given little guidance and direction to the tribes. As a result, a great deal of needless anxiety and apprehension have been created among traditional religious leaders, particularly the Hopis and Zunis. Many individuals have been left with the impression that they are going to have to disclose the locations of all of their sacred sites together with restricted sacred knowledge in order to protect these sites from possible destruction or to meet the terms of their contractual agreements. Most of the tribal studies are far broader and more comprehensive than are needed for protecting cultural resources from possible destruction caused by dam operations, and the actual risks associated with operations have remained poorly defined.

Only cultural resources that are located on the beaches and other areas along the river corridor need to be considered. Sacred knowledge concerning a particular location does not, and should not, need to be disclosed if it violates religious beliefs. For sites that can be identified, conditions need to be appraised and monitored over time, and methods need to be devised for protecting the cultural resources. The Hopis have devised an excellent minimal model for the type of site-specific data needed for protection (Dongoske, 1994).

Tribal studies should not be considered academic studies but rather applied studies focused toward specific objectives—that is, the protection of specific tribal cultural resources. Relevant to studies of sacred sites, in 1971 the Council of the American Anthropological Association adopted what it terms the "Principles of Professional Responsibility." These principles were amended in 1976 and again in 1989. Section 1 states:

> Anthropologists' first responsibility is to those whose lives and cultures they study. Should conflicts of interest arise, the interests of

> these people take precedence over other considerations. Anthropologists must do everything in their power to protect the dignity and privacy of the people with whom they work, conduct research or perform other professional activities. Their physical, social and emotional safety and welfare are the professional concerns of the anthropologists who have worked among them.

Few, if any, anthropologists engaged in field research have not at one time or another been privy to certain information the disclosure of which might prove detrimental to a particular individual or the community being studied. It is to be hoped that most anthropologists have withheld such information from public disclosure. There is little doubt that disclosure of restricted sacred knowledge would endanger the "social welfare" of tribal communities.

While restricted knowledge has been protected relative to the GCES, protection may be more difficult in the future. Studies show that in many cases two or more tribes claim the same cultural resource. Thus, jurisdictional disputes may arise between tribes over the control or protection of particular cultural resources. While the Navajo tribe notes some discussions with the Hopi and Hualapai researchers (Roberts et al., 1994), the tribes probably need to work together more extensively on their common concerns. It would be in the long-term interest of all of the tribes to cooperate with each other in the monitoring and protection of Native American cultural resources in the Grand Canyon.

SUMMARY

The tribes generally have not given explicit attention to the various flow alternatives or to how these alternatives might affect their cultural resources and values. This may have resulted from a lack of direction to the given researchers by the BOR. Some of the tribes, especially the Hualapai and Southern Paiute, have exhibited concerns for botanical resources but not so much for other biotic resources. All tribes have noted a concern for spirits in plants and animals. None of the reports, except that of the Navajo, indicate concern about the cost of electrical power. The Hualapai, especially with their river-running enterprise below Diamond Creek, are concerned with any effect river flows might have on recreational resources. The tribes have not addressed nonuse values in their reports except in a general way to the effect that the Grand Canyon has a sacred value whether they use it or not.

It is clear from anthropological, archaeological, and historical studies that Hopi, Southern Paiute, Hualapai, and Havasupai have all used the Grand Canyon to one degree or another in the past. The extent and significance of Navajo and Zuni occupation of the Canyon are as yet unclear.

RECOMMENDATIONS

1. The BOR should have involved the tribes in the Glen Canyon environmental research and on the cooperating group much earlier than it did. In future studies such as these, where Native American interests are apparent, the BOR should make sure the affected tribes are involved at the earliest possible planning stage.
2. In future such studies the BOR should provide more direction to the tribes involved so that they can more directly address the operation of the Glen Canyon Dam and its effects.
3. An anthropologist without a vested interest in any particular tribe or agency should be involved in future studies for which tribes participate in environmental and cultural research. An independent anthropologist could have enhanced the credibility of tribal reports and served as liaison to the tribes.
4. A determination should be made as to which archaeological sites are in danger of damage because of the operation of Glen Canyon Dam, and monitoring as well as mitigation need to be specified for the future.
5. The tribes should take responsibility for identifying sacred sites to the extent possible in terms of their individual religious precepts; the BOR and NPS should take responsibility to protect these sites.

REFERENCES

Bailey, G., and R. Bailey. 1986. A History of the Navajos: The Reservation Years. Santa Fe: School of American Research Press. Pp. 14-16.

Bunte, P.A., and R.J. Franklin. 1987. From the Sands to the Mountain: Change and Persistence in a Southern Paiute Community. Lincoln: University of Nebraska Press. Pp. 51-99, 240-241.

Burchett, T.W. 1993. Review Draft, Summary Report for the 1993 Glen Canyon Environmental Studies Monitoring of Archaeological Sites from Glen Canyon Dam to the Paria Riffle, Glen Canyon National Recreation Area. Flagstaff, Ariz.: Bureau of Reclamation.

Bureau of the Census. 1993a. 1990 Census of Population, Social and Economic Characteristics, Arizona. Washington, D.C.: U.S. Government Printing Office.

Bureau of the Census. 1993b. 1990 Census of Population, Social and Economic Characteristics, American Indian and Alaska Native Areas. Washington, D.C.: U.S. Government Printing Office.

Clemmer, R.O. 1995. Roads in the Sky: The Hopi Indians in a Century of Change. Boulder: Westview Press. Pp. 16, 274-295.

Dobyns, H.F. 1956. Prehistoric Indian Occupation Within the Eastern Area of the Yuman Complex. Unpublished master's thesis, Department of Anthropology, University of Arizona, Tucson.

Dobyns, H.F., and R.C. Euler, eds. 1961. The origin of the Pai tribes by Henry P. Ewing. The Kiva 26(Feb.):3.

Dobyns, H.F., and R.C. Euler. 1970. Wauba Yuma's People: The Comparative Sociopolitical Structure of the Pai Indians of Arizona. Prescott College Studies in Anthropology 3, Prescott College, Prescott, Arizona.

Dobyns, H.F., and R.C. Euler. 1974. Socio-Political Structure and the Ethnic Group Concept of the Pai. New York: Garland Publishing.

Dobyns, H.F., and R.C. Euler. 1976. The Walapai People. Phoenix: Indian Tribal Series.

Dongoske, K.E. 1993a. Progress Report Number 8 on the Hopi Tribe's Involvement as a Cooperating Agency in the Glen Canyon Dam Environmental Impact Statement. Kykotsmovi Village, Ariz.: The Hopi Tribe.

Dongoske, K.E. 1993b. Progress Report Number 9 on the Hopi Tribe's Involvement as a Cooperating Agency in the Glen Canyon Dam Environmental Impact Statement. Kykotsmovi Village, Ariz.: The Hopi Tribe.

Dongoske, K.E. 1994. Progress Report Number 14 on the Hopi Tribe's Involvement as a Cooperating Agency in the Glen Canyon Dam Environmental Impact Statement. Pp. 5-6. Kykotsmovi Village, Ariz.: The Hopi Tribe.

Euler, R.C. 1958. Walapai culture history. Ph.D. dissertation, Department of Anthropology, University of New Mexico, Albuquerque.

Euler, R.C. 1966. Southern Paiute Ethnohistory. University of Utah Anthropological Papers No. 78, Salt Lake City.

Euler, R.C. 1972. The Paiute People. Phoenix: Indian Tribal Series.

Euler, R.C. 1974. Havasupai Historical Data. New York: Garland Publishing.

Euler, R.C. 1981. Cohonina-Havasupai relationships in Grand Canyon. In Collected Papers in Honor of Erik Kellerman Reed, Albert H. Schroeder, ed. Papers of the Archaeological Society of New Mexico, vol. 6. Albuquerque: Albuquerque Archaeological Society Press.

Fairley, H.C., P.W. Bungart, C.M. Coder, J. Huffman, T.L. Samples, and J.R. Balsom. 1994. The Grand Canyon River Corridor Survey Project: Archaeological Survey Along the Colorado River Between Glen Canyon Dam and Separation Canyon. Prepared in cooperation with the Glen Canyon Environmental Studies. Arizona: Grand Canyon National Park.

Ferguson, T.J., and E.R. Hart. 1985. A Zuni Atlas. Norman: University of Oklahoma Press. Pp. 20-23, 44-51.

Heib, L.A. 1979. Hopi World View. Handbook of North American Indians, Southwest, vol. 9. Washington, D.C.: Smithsonian Institution Press. Pp. 577-580.

Hereford, R. 1993. Description of Map Units and Discussion to Accompany Map Showing Surficial Geology and Geomorphology of the Palisades Creek Archaeologic Area, Grand Canyon National Park, Arizona. USGS Open-File Report 93-553. U.S. Geological Survey, Reston, Va.

Hereford, R., H.C. Fairley, K.S. Thompson, and J.R. Balsom. 1993. Surficial Geology, Geomorphology, and Erosion of Archaeologic Sites Along the Colorado River, Eastern Grand Canyon, Grand Canyon National Park, Arizona. USGS Open-File Report 93-517, U.S. Geological Survey, Reston, Va.

Hopi Tribe. 1990. Proposal for Involvement of the Hopi Tribe in GCES Projects in the Colorado River Corridor and Little Colorado River. Kykotsmovi Village, Ariz.: The Hopi Tribe.

Hualapai Cultural Resources Division. 1993. Hualapai Tribe Ethnographic and Oral Historical Survey for Glen Canyon Environmental Studies and the Glen Canyon Dam Environmental Impact Statement. Peach Springs, Ariz.: Hualapai Tribe.

Johnston, D.F. 1966. An Analysis of Sources of Information on the Population of the Navajo. Smithsonian Institution, Bureau of American Ethnology, Bulletin 197. Washington, D.C.: U.S. Government Printing Office. Pp. 136-137, 153.

Jones, A.T., and R.C. Euler. 1979. A Sketch of Grand Canyon Prehistory. Grand Canyon: Grand Canyon Natural History Association.

Kappler, C.J. 1904. Indian Affairs: Laws and Treaties, vol. 1. Washington, D.C.: U.S. Government Printing Office. Pp. 176-177, 804-809.

Kelly, I.T. 1934. Southern Paiute bands. American Anthropologist 36:4.

Kelly, I.T. 1964. Southern Paiute ethnography. University of Utah Anthropological Papers No. 69, Salt Lake City.

Reichard, G.A. 1928. Social Life of the Navajo Indians: With Some Attention to Minor Ceremonies. New York: Columbia University Press. Pp. 11-19.

Roberts, A., R. Begay, and K.B. Kelley. 1994. Bits'iis Nineezi (The River of Neverending Life): Navajo History and Cultural Resources of the Grand Canyon and the Colorado River. Window Rock: Navajo Nation Historic Preservation Department. Pp. vii, 3.

Schwartz, D.W. 1989. On the Edge of Splendor: Exploring Grand Canyon's Human Past. Santa Fe: School of American Research.

Stoffle, R.W., D.B. Halmo, M.J. Evans, and D.E. Austin. 1993. Pia'paxa'huipi (Big River Canyon): Ethnographic Resource Inventory and Assessment for Colorado River Corridor, Glen Canyon National Recreation Area, Utah and Arizona, and Grand Canyon National Park, Arizona. Presentation paper, June 24. Pp. 1-3.

Stoffle, R.W., L.L. Loendorf, D. Austin, D. Halmo, A.S. Bulletts, and B.K. Fulfrost. 1995. Tumpituxwinap: Southern Paiute Rock Art in the Colorado River Corridor. Preliminary draft, January. Pp. 87-231.

Thomas, J.R. 1993. Navajo Nation Position Paper, Glen Canyon Dam Environmental Impact Statement. Window Rock: The Navajo Nation. Pp. 15-16.

U.S. Department of the Interior. 1989. Notice of final determination that the San Juan Southern Paiute Tribe exists as an Indian tribe. Federal Register 54(240):1502-1505.

Vogt, E.Z. 1961. Navaho. Pp. 278-336 in Perspectives in American Indian Culture Change, E. H. Spicer, ed. Chicago: University of Chicago Press.

Williams, A.W., Jr. 1970. Navajo Political Process. Smithsonian Contribution to Anthropology, vol. 9. Washington, D.C.: Smithsonian Institution Press. P. 13.

Woodbury, R.B. 1979. Zuni Prehistory and History to 1850. Handbook of North American Indians, Southwest, Vol. 9. Washington, D.C.: Smithsonian Institution Press. Pp. 467-473.

9

Power Economics

INTRODUCTION

Power resources occupied an important and somewhat unique position in the Glen Canyon Environmental Studies (GCES). Electrical output from Glen Canyon is proportional to water flow through the dam, and the value of that output varies daily, weekly, and seasonally. As a result, power economics and water flows are closely related.

Traditionally, Glen Canyon Dam has operated with relatively few constraints so as to maximize the value of its electrical output (Chapter 4). Thus, operational changes (such as those implemented under the interim flow regime and those that were under active consideration in the environmental impact statement (EIS) for Glen Canyon Dam) alter the scheduling and reduce the value of power production. This loss of power resources accounts for most, if not all, of the costs of altered dam operations.

As can be seen from the previous chapters, operational changes at Glen Canyon Dam can have beneficial effects on native fishes, beaches, recreation, and archeological sites. Therefore, a principal focus of the decision making process concerning operation of the dam is the balance between the value of production of electricity and other resources. This has several important implications. First, since changes to dam operations reduce the value of power, it is those who benefit from this output who will tend to be affected most adversely and thus to be opposed to such changes. Interests related to power production are typically well defined, strongly organized, and quite aggressive in advocating their point of view. By contrast, those who might benefit from modified dam operations are more diverse and must be motivated by somewhat more diffuse goals, involving environmental or cultural resources.

Second, the contrasting position of the different interests is accentuated by the nature of the resources. Electrical output can be measured readily. Moreover, it is a good that is bought and sold and that can be assessed a specific monetary value. Other resources are generally harder to measure, and it can be difficult or impossible to assign them monetary values (Chapter 7). Thus, in the EIS for Glen Canyon Dam and elsewhere, the costs of changes to dam operations are reported in dollars, while most of the benefits are reported in other units (e.g., number of beaches).

Third, the disjunction between electrical power and other resources is further accentuated by the contrasting levels of historical analysis. Power resources have been the subject of decades of analysis. The utilities potentially affected by changes in operations at Glen Canyon Dam are commonly viewed as having sufficient data and expertise to estimate the adverse effects on their interests. Prior to the GCES, natural resources were subject to much less study. Without the requisite data and expertise, it was difficult to assess how these resources would be affected by changes in dam operations. The GCES increased the feasibility of comparisons between power production and the natural resources of Glen Canyon.

The GCES has also played a major role in advancing the study of Glen Canyon Dam power economics. It became clear during Phase I of GCES that the existing analysis did not provide an adequate basis for decision making regarding altered dam operations. The quality of this analysis has substantially improved as a result of the extensive power economics studies undertaken during GCES Phase II. This process has greatly benefited from broader public participation and review external to federal agencies and utilities.

FLOWS AFFECT ELECTRICAL OUTPUT AND COSTS

The Colorado River Storage Project Act directs the Secretary of the Interior to operate power plants "so as to produce the greatest practicable amount of power and energy that can be sold at firm power and energy rates." In a hydro project such as Glen Canyon, water is impounded behind a dam and discharged at a lower elevation into the river downstream. The mechanical energy of the falling water is used to turn turbines, which generate electricity.

Electrical output is measured in two ways. Instantaneous output is referred to as power and is measured in terms of watts (W). Output over time is electrical energy and is expressed in terms of watt-hours (Wh). The quantity of electrical output is typically much larger than a watt, so units such as kW (thousand watts), MW (million watts), and GW (billion watts) are used. Similarly, typical units for energy are kWh, MWh, and GWh.

The maximum amount of power that can be produced by a power plant is known as capacity. For a hydro dam, capacity is a function of the number and size of turbines. Subsequent to a rewinding and uprating of generators completed in 1987, Glen Canyon Dam capacity has been 1356 MW (at 33,200 cubic feet per second (cfs), which is the maximum flow through the turbines) (BOR, 1995). However, the Bureau of Reclamation (BOR) agreed not to use this increased capacity pending completion of a comprehensive study of the effects of historic and current dam operations on environmental resources. Thus, Glen Canyon Dam has generally been limited to a capacity of 1,300 MW (at 31,500 cfs of flow) (PRC, 1995).

The maximum amount of energy that can be produced over time by a hydro dam is determined by the smaller of two constraints: turbine capacity and amount of water in the reservoir. If Glen Canyon Dam were to operate at full capacity continuously for 1 year, it would generate almost 12,000 GWh and discharge 24 million acre-feet (maf) of water. This greatly exceeds the amount of water entering Lake Powell, even in wet years (Chapter 4). At the average annual flow of 10 maf, Glen Canyon Dam can produce about 5,000 GWh annually.

Because turbine capacity generally exceeds water supply, the annual electrical output of a dam is determined by the amount of water in the reservoir. Thus, changes in daily or monthly operations will have no effect on annual power generation (Chapter 4). They will, however, affect the scheduling of electrical output and thus its value.

The value of electricity varies substantially with time. Demand for electricity fluctuates daily, weekly, and seasonally. It is higher during the day, when businesses are open, and lower at night and on weekends. It is greater during the winter and summer, when required for heating and cooling, than during the fall or spring. In addition, electricity cannot be cheaply or conveniently stored.

The reliability of the electrical power system is also difficult to maintain. Power plants and customers throughout western North America are interconnected, and electricity moves at the speed of light. If generation and consumption are not continuously and closely matched, the power system becomes unstable and blackouts can result. Thus, utility systems are planned to have sufficient generating capacity to supply customer requirements and provide a reserve for malfunctions and other exigencies.

The cost of producing electricity includes two principal components: (1) fixed costs to build power plants and keep them in operable condition and (2) variable costs associated with operation. For fossil fuel power plants, the variable costs relate mostly to fuel purchases. For hydro plants, which are powered by water, variable costs are relatively low, but the fixed costs of dam construction are high.

In a typical large power system, several kinds of generating plants are used. In general, plants with high fixed costs and low operating costs (such as coal-fired stations) serve the base load, while plants with low fixed costs and higher operating costs (such as gas- or oil-fired stations) are used to meet peak demand. Overall, base load is cheaper to serve than peak demand because fixed costs can be spread over more hours of output.

One notable advantage of hydro plants is that they can respond quickly to variations in demand for electricity. Hydro turbines can be turned on and off almost instantaneously. In contrast, conventional thermal powerplants use boilers to generate steam for a turbine and can require substantial start-up time (ranging from minutes to hours) to generate electricity. Thus, hydro plants are especially well suited for providing peaking power.

With the large turbine capacity at Glen Canyon Dam, traditional operations provided great flexibility to schedule electrical production at times when it would be most valuable. Changes in dam operations have restricted the maximum flows available during peak periods. These changes have the effect of shifting power output from periods when it is more valuable to periods when it is less valuable, with essentially no change in annual energy production. With less output during peak periods, additional supply is required from other sources. Also, a quantity of off-peak power from Glen Canyon Dam is available for sale to other utilities or to displace the need for other power supplies. Thus, the cost of altered flows has been the difference between (1) the cost of peak power required to replace the output shifted to off-peak periods and (2) the value of this incremental off-peak power.

THE INSTITUTIONAL CONTEXT

Glen Canyon Dam is owned and operated by the Bureau of Reclamation. The Western Area Power Administration (WAPA) markets this electricity on a wholesale basis to about 180 preferred customers. As required by federal law, these preferred customers are municipal and county utilities, rural electric cooperatives, water and irrigation districts, U.S. government installations, and other non profit organizations. Each individual customer serves a designated area in the region. As shown in Figure 9.1, most are in the six states of Arizona, Colorado, Nevada, New Mexico, Utah, and Wyoming, although some extend into California, Nebraska, and Texas. These preferred customers, in turn, serve 1.7 million end-use customers, including residential, commercial, industrial, and agricultural users.

In part, the distribution of costs associated with altered flows is determined by the contractual arrangements for sale of Glen Canyon Dam electricity. Under existing contracts, WAPA's costs will increase, because it is obligated to supply fixed quantities of peaking power, which it may have to purchase at higher costs from other utilities. As contracts expire and are renegotiated, however, WAPA could contract to sell less firm peak power and more firm off-peak power. In this case, both the cost and the value of WAPA's electricity will diminish, and the need to replace the dam's output will be shifted to WAPA's customers. Thus, there is a relationship between these marketing policy issues and the costs associated with altered flow regime. A separate EIS has been established to address marketing issues (WAPA, 1994).

It is important to distinguish between the *economic* perspective that measures changes in overall costs to society and the *financial* perspective concerning changes in the costs borne by specific entities. In the short term, economic costs caused by altered flows will be limited, because the region currently enjoys a surplus of generating capacity (in excess of the required reserve margin). Under the national economic perspective, the fixed costs of this existing capacity are excluded because they must be paid whether or not the capacity is used as a replacement for Glen Canyon Dam.

From a financial point of view, however, there will be differences in costs and benefits among utilities. Utilities that have surplus capacity and can sell power to replace Glen Canyon Dam stand to gain at the expense of other utilities that must purchase power to replace Glen Canyon Dam power. Surpluses eventually will decrease, and prices for peak power can be expected to rise. Eventually, new supplies will be required, which will entail both economic costs to society and financial costs to utilities.

Within the restricted six-state marketing area for Glen Canyon Dam, 70 percent of electricity consumers (3.9 million) are served by utilities that do not receive power from the dam. Absent some new type of cost-sharing mech-

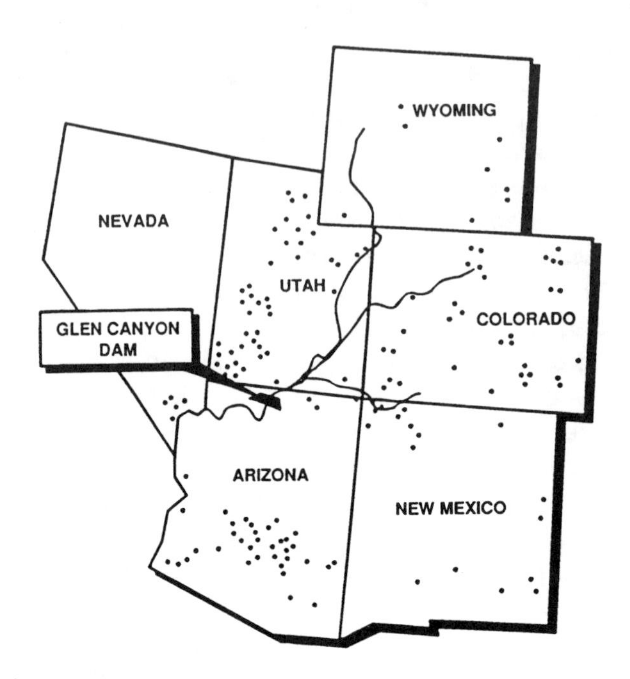

FIGURE 9.1 The Salt Lake City Area/Integrated Projects markets power to approximately 180 utilities, mostly in six states. SOURCE: Bureau of Reclamation (1995).

anism, electricity bills for these utilities will be unaffected by changes at Glen Canyon Dam, or may decrease owing to surplus power sales. Notably, most of the region is served by investor-owned utilities (e.g., Arizona Public Service, Nevada Power, Public Service of Colorado, Utah Power), which are foreclosed by law from being WAPA preferred customers. Six relatively large utilities, which serve 1.3 million end-use customers, receive about half of Glen Canyon Dam's output. The remainder goes to numerous smaller systems, which together supply only 400,000 customers (7 percent of the region's total).

Glen Canyon Dam operations, and the resulting revenues, also relate to a broader set of issues. Access to the relatively low cost electricity produced by the dam is restricted; about half goes to only 7 percent of regional electricity consumers. Recipients of the dam's electricity benefit from the use of public water resources that were developed with low-cost government debt. Under the terms of the Colorado River Storage Project Act, revenue from Glen Canyon Dam is intended to support reclamation irrigation projects.

Judgments about the appropriateness of these historical arrangements influence determinations of how the costs of altered flows should be distributed. If the beneficiaries of Glen Canyon Dam have traditionally been subsidized at the expense of taxpayers and the environment, it is acceptable that they bear the costs of altered operations. Not surprisingly, this point of view is strenuously resisted by the beneficiaries of the dam, notably the Colorado River Energy Distributors Association (CREDA), which represents the utilities receiving the bulk of Colorado River hydropower. They assert that the beneficiaries of Glen Canyon Dam have paid their fair share and should be shielded from any increased costs (Barrett, 1992a, 1992b).

COST ESTIMATES FOR ALTERED FLOW REGIMES

Analyzing the costs of altered flow regimes is complex. The operation of Glen Canyon Dam must be simulated to provide a realistic estimate of the time pattern of power production under the various constraints of each flow regime. These simulations must be carried out over a number of years to reflect annual variations in hydrology.

Once the output of the dam under different flow regimes has been established, the costs of these regimes to power users can be estimated. Such analyses typically rely on computer models, which are used to simulate the planning and operation of electric utility systems. These models require extensive data on existing power plants, new supply options, fuel costs, required reserve margins, interconnections between utilities, and current and projected electricity demand. They use various mathematical techniques to determine the optimal (least-cost) solution, subject to specific constraints. Output includes data on the operation of each power plant, the new power plants and demand-side management (conservation) programs that are used to meet requirements for new supply, and capital and operating costs.

Relatively little work regarding power economics was undertaken during Phase I of GCES. In response to the National Research Council (NRC) committee's review of GCES Phase I (NRC, 1991), WAPA was requested to analyze the economic effects of increasing minimum flows to either 5,000 or 8,000 cfs all year. WAPA estimated annual costs of $5.0 and $14.7 million, respectively, during the 1990s and much larger costs subsequently (NRC, 1991). These cost estimates were based solely on a WAPA financial perspective. Moreover, the modeling of Glen Canyon Dam operations did not provide a realistic estimate of the time pattern of power production under the constraints of each flow regime (NRC, 1991). Therefore, the loss of peak period capacity and energy output due to altered flows was overstated.

As part of GCES Phase II, detailed power resource studies were initiated in November 1988. These studies were conducted by the Power Resources Committee (PRC). Three of the four members have historically controlled and benefited from Glen Canyon Dam power resources: BOR, WAPA, and CREDA. In addition, however, the PRC included the Environmental Defense Fund (EDF), a national environmental group. As the developer of a widely used utility simulation model (ELFIN, or Electric Utility Financial and Production Cost Model), EDF provided the PRC with substantial technical expertise regarding utility economics, as well as an alternative point of view.

The PRC's work focused on Glen Canyon Dam. However, during the early part of GCES Phase II, individual members of the PRC also produced some analyses of costs related to experimental and interim flows. WAPA estimated overall costs of $10.9 million for experimental flows planned for 1990 and 1991 (NRC, 1991). Once again, these cost estimates were based solely on a WAPA financial perspective. The modeling of Glen Canyon Dam operations was an improvement over previous work but still did not provide a realistic estimate of the time pattern of power production under the constraints of each flow regime (NRC, 1991).

EDF used ELFIN to evaluate the interim flow regimes proposed by various parties, including WAPA (EDF, 1991). These estimates were based solely on a national economic perspective, which EDF argued should be given preference over the financial perspective of affected utilities. Estimated costs for WAPA's proposal were only $1 million to $2 million annually over the period 1992 to 1995. For the other proposals, the costs were on the order of $9 million for 1992, increasing to $15 million to $16 million for 1995.

For interim flows the BOR's draft environmental assessment noted that conclusive data were not yet available from the detailed power resource studies under way as part of GCES Phase II (BOR, 1991). The costs were expected to be small, however, for the 3-year period of interim flow, because surplus capacity was likely to be available. From a national economic perspective, effects were characterized as a minor increase. The draft environmental assessment stated that the financial effects on WAPA were estimated to be $22 million in fiscal year 1992. If financial exception criteria were provided, allowing limited exceedances flow criteria, costs would be reduced to only $3 million.

Exception criteria were ultimately accepted by the BOR, and the costs of interim flows have been relatively low. To date, WAPA has avoided the need to purchase replacement capacity and has been able to operate with 10 to 15 percent available capacity above peak needs, as opposed to 30 percent previously (BOR, 1995).

The efforts of the PRC were first directed at determining the best approach to analyzing power economics. A detailed study (BOR, 1990) recommended that the EGEAS (Electric Generation Expansion Analysis System) and ELFIN simulation models be used and their results compared. EGEAS has the capability to perform expansion planning (addition of new resources) as well as production cost estimates (dispatch of a given set of resources). ELFIN estimates production costs and utility financial models.

For the power resources studies, EGEAS modeling was carried out by Stone and Webster Consultants, Inc., a contractor retained by the BOR through HBRS, Inc. (the contractor for GCES recreation and nonuse value economics studies). The expansion plans developed by using EGEAS were used as input to the ELFIN modeling conducted by EDF. The production cost projections of the two models were similar.

Power system costs were measured over a 50-year period. To allow for changes at Glen Canyon Dam to influence the need for new supplies, the most detailed analysis was conducted for an initial 20-year planning period (1991-2010). To reflect costs over the lifetime of power plants, a 30-year extension period (2011-2040) was modeled. Continued escalation in fuel and other costs was assumed, but the level of demand and set of supply resources forecast for 2010 were held constant (resources being retired were assumed to be replaced in kind).

The operation of Glen Canyon Dam also had to be simulated to estimate the time pattern of power production under each flow regime. The BOR's Colorado River Simulation Model (CRSM) was the source of projections for monthly water releases, capacity, and energy. CRSM has no capability for simulating hourly operational constraints, such as those imposed by fluctuating flow alternatives. As discussed earlier, two marketing approaches were modeled. For the Contract Rate of Delivery (CROD) marketing approach, a simple geometric method was used to determine available capacity. The peak shaving algorithm from ELFIN was utilized for the hydrology marketing approach.

The foregoing description only begins to capture the intricate array of analyses undertaken for the power resources studies (see Figure 9.2). In large part, this complexity results from efforts to reflect the institutional context, with its multiplicity of affected parties, viewpoints, and practices. In particular, much of the effort in the power resources studies was devoted to

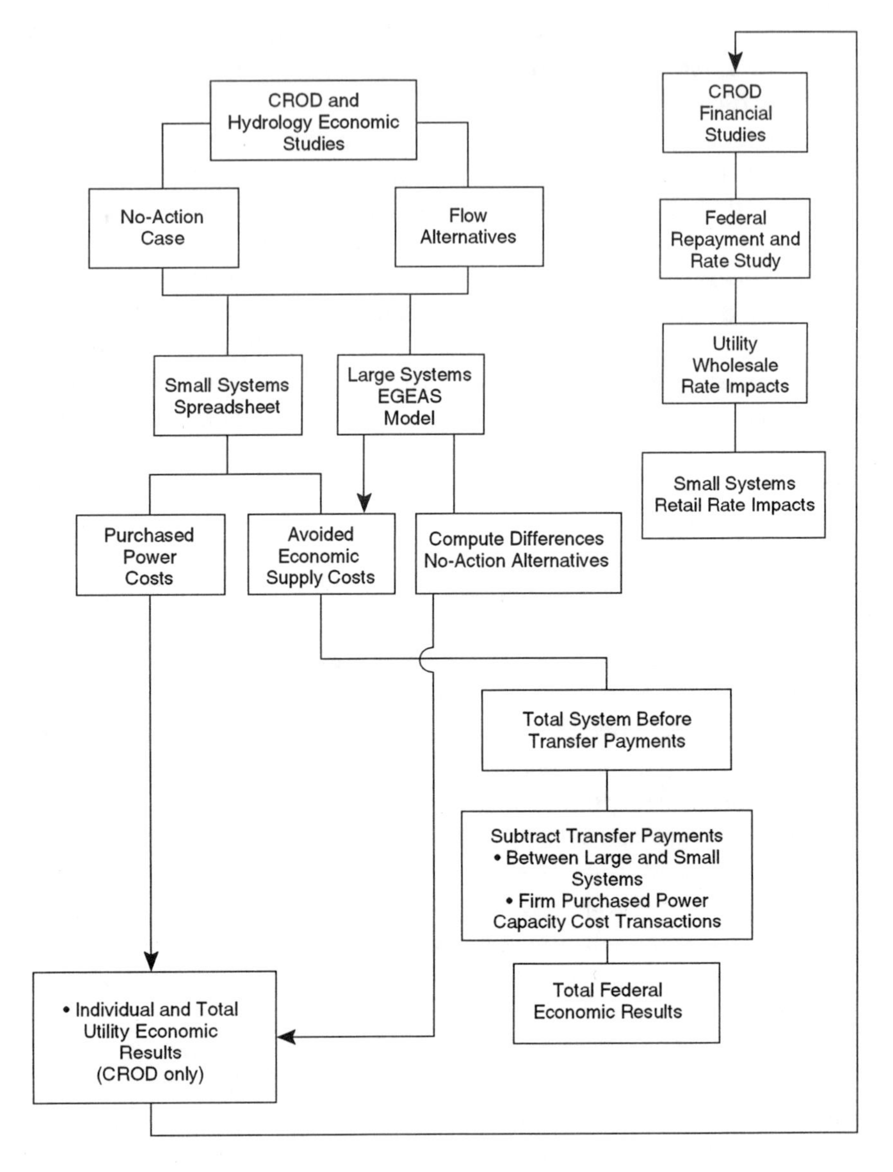

FIGURE 9.2 Study process to determine power values. SOURCE: Adapted from Power Resources Committee (1994, Fig. I-3).

detailed modeling of the financial and rate effects on individual utilities.

Even for the analysis of the national economic perspective, separate simulations were conducted for each of seven large utilities (one of these utilities has now been absorbed by the others), with some attempt to reconcile and coordinate the individual analyses. Meanwhile, the small utilities, which rely on WAPA and other utilities for most or all of their supply, were not included in the EGEAS modeling. These were assumed to purchase replacement power from their alternative suppliers, at costs based on the EGEAS model of large systems.

It would have been simpler to model all of the utilities as a single integrated system. This approach was recommended by the NRC committee and rejected by the PRC (NRC, 1992; Roluti, 1993; Power Resources Committee, 1993, 1994). The rationale was that the individual utilities do not now coordinate the operation of their systems so as to minimize overall costs. Moreover, existing transmission capacity between utilities is limited. Thus, results based on a single optimized system could understate the actual costs of system operations. However, even within the constraints imposed, the resource plans selected were not completely optimized (Power Resources Committee, 1993, 1994).

The focus on individual utilities and financial effects also necessitated detailed modeling of WAPA's marketing practices. The data on Glen Canyon Dam's power production had to be combined with estimates of the output from other hydro projects marketed by WAPA to determine the amount of power and energy that would be contracted for sale to preferred customers. This determination was affected by the marketing criteria used by WAPA. Currently, WAPA uses the approach, CROD (Contract Rate of Delivery) according to which the amount of firm capacity and the amount of energy are fixed in advance, and WAPA must purchase electricity to supplement hydro output in dry years. Alternatively, under the *hydrology* approach, WAPA would sell only the capacity and energy available given actual hydro output. Customers, rather than WAPA, would be responsible for meeting any additional needs. The power resources studies modeled both the CROD and hydrology approaches.

Extensive analyses were performed of the sensitivity of results to changes in input assumptions or methodology. Sensitivity analyses are standard practice in power economics studies. In keeping with recent trends, the Glen Canyon Dam analyses relied to some extent on sophisticated probabilistic approaches. The use of sensitivity analyses, however, is not a substitute for selection of appropriate base-case assumptions. When the PRC was unable to reach consensus on these assumptions, remaining disagreements were typically "resolved" by specifying a case for sensitivity analysis. This was also the response to certain concerns expressed by the NRC committee (NRC, 1992; Roluti, 1993; Power Resources Committee, 1993, 1994).

The EGEAS model was also used to estimate emissions of atmospheric pollutants. Emissions rates for each power plant (tons of sulfur dioxide (SO_2) and nitrogen oxides (NO_x) per unit of fuel burned) were provided as input. The model then combined these data with the results concerning power plant operations to estimate the total tons of SO_2 and NO_x.

Review of the power resources studies is further complicated by the significant changes that occurred over the course of GCES work. The PRC produced three reports that were provided to the NRC committee: a draft Phase II report (PRC, 1992), a final Phase II report (PRC, 1993), and a Phase III report (PRC, 1995). Each report incorporated improvements in methodology and data from the previous one. These improvements addressed problems in the analysis identified by the NRC committee and other reviewers (NRC, 1992; NRC, 1994).

Unfortunately, the usefulness of GCES Phase II and Phase III results is compromised by changes in the flow alternatives that were modeled. The Phase II studies analyzed eight flow alternatives (including "no action"). Prior to the release of the draft EIS, beach-building and habitat maintenance flows were added to the moderate fluctuating flow and seasonally adjusted steady flow alternatives; they were also included in the new preferred alternative of the EIS (modified low fluctuating flows). The Phase II studies were too advanced to incorporate these changes, so consideration of them was deferred to Phase III.

Because of resource constraints, Phase III work was limited to the "no action" and preferred alternatives. As a result, there are no Phase III results for most flow alternatives, and the final EIS largely relies on the earlier Phase II work. For the three flow alternatives that include beach-building and habitat maintenance flows, cost data for the final EIS were derived from a simple regression analysis based on the amount of capacity lost and the costs estimated for Phase II flow alternatives. The final EIS does provide the Phase III results as supplemental data for the preferred alternative.

Table 9.1 summarizes the most important numerical results from the power resources studies. In the interest of brevity, data are presented for only three flow regimes. "No action" serves as the benchmark for measuring the effects of altered flows. Seasonally adjusted steady flows result in the greatest power resource costs and potentially the greatest benefits in terms of protecting other resources. The preferred alternative and seasonally adjusted steady flow bracket the range of operational changes likely to be im-

TABLE 9.1 Summary of the Effects of Three Operating Regimes on the Value of Electrical Power From Glen Canyon Dam

	No Action	Preferred Alternative: Modified Low Fluctuating Flow		Seasonally Adjusted Steady Flow
Marketable Resource				
Annual Energy (GWh)	6,010	6,018 (+0.1%)		6,123 (+1.9%)
Winter Capacity (MW)	1,407	965 (-31.4%)		640 (-54.5%)
Summer Capacity	1,315	845 (-35.2%)		498 (-62.1%)
Increase in Economic Costs (relative to No Action)				
Annual (nominal $ million)		Phase II	Phase III	
CROD	$0	$44.2	$34.8	$123.5
Hydrology	$0	$15.1	$25.0	$88.3
Present Value (1991 $ million)				
CROD	$0	$511.2	$402.0	$1,428.4
Hydrology	$0	$174.6	$286.8	$1,021.2
WAPA Wholesale Rate (FY 1993 ¢/kWh)	1.878	2.316 (+23.3%)		2.820 (+50.2%)
Retail Rates (¢/kWh)				
Customers				
Large systems (23% end users)	No change	Slight decrease to moderate increase	See text p. 178	Slight decrease to moderate increase
Small systems (7% end users)	6.41	7.05 (+10.0%)	See text p. 178	7.58 (+16.3%)
Other regional utilities (70% of end users)	No change	No change to slight decrease		No change to slight decrease

plemented and that might occur in connection with endangered-fish research. Costs for fluctuating flow regimes not shown in Table 9.1 are generally intermediate between those for "no action" and the preferred alternative, while those for other steady flow regimes are intermediate between the preferred alternative and seasonally adjusted steady flow.

The preferred alternative reduces WAPA's marketable capacity by approximately 450 MW. In the power resources studies, the stream of annual nominal dollar costs is present valued to 1991 using an 8.5 percent *nominal* discount rate. Based on the GCES Phase III study, the present value of the associated economic cost is on the order of $300 to $400 million. Levelized over the 50 year analysis period, this is equivalent to an annual cost of $25 to $35 million in nominal dollars or $15 to $20 million in 1991 dollars.

The power resources studies report annual costs on a nominal levelized basis; over the 50-year period being analyzed, the estimated equivalent annual effects remain constant in nominal terms. In constant (inflation-adjusted) dollars, they are highest in the first year and decline steadily due to inflation. The figures for 1991 dollar levelized costs in this report were calculated based on the *real* discount rate. Given the 3.8 percent inflation rate assumed (Power Resources Committee, 1993, p. III-13), the 8.5 percent nominal discount rate is equivalent to a 4.5 percent real discount rate. For a 50 year period, this yields a real annualization factor of 5.08 percent. Annual impacts in 1991 dollars are approximately 41 percent less than those calculated on a nominal basis.

Costs incurred in any given year are expected to vary substantially over time. In general, they would be lower in earlier years. The region has a surplus capacity, which the power resources modeling assumes would be prolonged as the large utilities implement demand-side management programs. Costs would rise after 1998 as new capacity is required. Costs will also be affected by variations in hydrology. The constraints associated with altered flows will have less effect during wet years.

For the preferred alternative, WAPA's wholesale rates are estimated to increase by about 0.5¢/per kilowatt-hour, but Glen Canyon Dam power would still be highly competitive with alternative sources. Most regional electricity customers would experience little, if any, change in their retail rates. For several reasons, however, the effect of changes in operations at Glen Canyon Dam will tend to be much greater for the small preferred customers. They typically buy most or all of their power from other utilities and rely heavily on Glen Canyon Dam power. By contrast, the large preferred customers generate much of their own power. They currently have surplus capacity, which can be used to replace lost Glen Canyon Dam power for their own needs and through expanded sales to the small systems.

The final EIS for Glen Canyon Dam reports that increases in small system retail rates would range from 4 to 16 percent, averaging 10 percent or about 0.6¢/per kilowatt-hour. The basis of these estimates is unclear. The draft EIS reported smaller effects on rates, similar to those shown in the GCES Phase II study (PRC, 1993). The final EIS cites the Phase III study (PRC, 1995), but the version of the report provided to the NRC committee does not include the final EIS data. The data that are provided in the Phase III report indicate that effects on rates will vary both across utilities and over time, with a *maximum* increase of 9.5 percent for one utility in 1 year. On average over the 1991-2010 period, rates will rise by 1 percent or less for Rural Electric Administration member utilities and by 2 to 5 percent for municipal utilities.

Seasonally adjusted steady flow reduces WAPA's marketable capacity by approximately 800 MW. The present value of the associated economic cost is estimated to be on the order of $1 billion to $1.4 billion. Levelized over the 50-year analysis period, this is equivalent to an annual cost of about $90 to $120 million in nominal dollars or about $50 to $70 million in 1991 dollars. As for the preferred alternative, costs would generally be lower in the earlier years. However, the greater loss of capacity would advance the need for construction of new capacity and the associated costs. Cost estimates based on Phase III methodology would likely be lower than these, which are based on the Phase II approach, which does not value off-peak energy correctly.

WAPA's wholesale rates are estimated to increase by about 1¢/per kilowatt-hour. Even so, Glen Canyon Dam power would still be competitive with all but the lowest-cost alternative sources. Most regional electricity customers would experience limited, if any, change in their retail rates. Reported increases in small system retail rates would range from 8 to 33 percent or, on average, 18 percent or about 1.2¢/per kilowatt-hours.

In summary, the cost effects of the preferred alternative are relatively modest. Costs for seasonally adjusted steady flows are generally two to three times greater than those for the preferred alternative. Even costs of this magnitude, however, would have only limited effects on agriculture and no material impact on the overall regional economy (BOR, 1995; WAPA, 1994; Flaim et al., 1994). For any altered flow regime, some small utility customers may bear a disproportionate share of the costs; however, they also received a disproportionate share of the benefits of low-cost Glen Canyon Dam electricity in previous years.

The costs of altered flows may be less than estimated, especially for the small utilities, which account for most of these costs. The GCES power resources studies did not consider the option of having WAPA maintain the same marketing commitment and use its transmission system to procure low-cost replacement capacity on behalf of its customers. Studies in support of the marketing EIS indicated that this approach could considerably reduce economic and financial effects (BOR, 1995). Even without such an approach, changes in the electricity industry (e.g., the U.S. Energy Policy Act of 1992) are providing greater access to a wide variety of low cost electricity supply sources, especially for small utilities, which traditionally have been limited in their supply options.

The base-case analysis in the power resources studies assumes a rapid escalation in oil and gas prices (averaging 8.4 percent nominal or 4.4 percent real annually), which now appears highly unlikely. Lower oil and gas prices would reduce the costs of operating the peaking plants used to replace lost capacity at Glen Canyon Dam. The power resources studies also did not explicitly consider the relationship between electricity prices and the amount of energy consumed (price elasticity). If rates increase because of altered flows, this will reduce future electricity demand. In turn, this will delay the need for new capacity and reduce the cost of altered flows. The PRC chose to deal with these issues through sensitivity analyses, which confirm that the estimated costs of altered flows will be substantially lower if lower fuel prices and lower demand materialize in the future.

The costs associated with altered flows must be compared with the benefits. The preferred alternative reduces regional SO_2 and NO_x power plant emissions by almost 1 percent, as fossil fuel generation is shifted from base load to cleaner peaking plants and construction of new cleaner plants is advanced. Under the Clean Air Act Amendments of 1990, there is a market for SO_2 emissions allowances. On this basis, the GCES Phase III report estimates that SO_2 emissions reductions are worth about $5 million (1991 net present value) or about 1 to 2 percent of the estimated increase in electricity costs (PRC, 1995). Seasonally adjusted steady flow would likely result in even greater reductions in emissions. The power resources studies did not consider other types of environmental impacts, such as those associated with fuel supply and transportation. Nonetheless, it is reasonable to assume that by shifting electricity production to newer and cleaner power plants, altered flows will generally reduce adverse environmental impacts.

The preferred alternative yields increased recreation values of $43.3 million (1991 present value), equivalent to about 10 to 15 percent of the estimated increase in electricity costs (BOR, 1995). Nonuse values are of much greater magnitude than the increase in power costs (Chapter 7).

RECOMMENDATIONS

The GCES power resources studies were impeded by an unfortunate combination of factors. The process was dominated by the entities that historically have controlled and benefited from Glen Canyon Dam power resources, notably BOR, WAPA, and CREDA. These entities have a clear incentive to deter implementation of altered flows, which would reduce the value of the dam's electrical output. On the other hand, they have the expertise to perform power resource studies, in light of their familiarity with these issues and high level of analytical resources. In fact, however, very little useful information regarding the cost effects of altered flow regimes was provided by those entities during Phase I of GCES.

During GCES Phase II there was great progress in developing the requisite tools for measuring cost effects. Nonetheless, the end result has not been wholly satisfactory in terms of providing cost estimates that are accurate, well documented, and readily reviewable. Unfortunately, the data in the EIS are principally based on Phase II Power Resources studies, rather than the subsequent Phase III analyses, which generally indicate lower costs for altered flows.

It is difficult, even with hindsight, to make completely definitive judgments of many aspects of the power resources studies. To some degree, the multiple analyses that were undertaken reflect the complex institutional context and the distribution of costs and benefits across different groups. Given the constraints in terms of budget and schedule, however, the strong focus on distributional issues has adversely affected the accuracy and timeliness of the analysis from a national economic perspective. Given that Glen Canyon Dam is federally owned and affects resources of national (and international) significance, the national economic perspective should be given precedence in the future. This approach is in keeping with the principles and guidelines established for federal water projects (Water Resources Council, 1983).

The power resources studies generally assumed that current practices and constraints would remain in place throughout the 50-year analysis period. This is problematic because the electric utility industry is evolving toward a more competitive future, which should help reduce the cost of altered flow regimes. The PRC missed a valuable opportunity to inform decision makers concerning the effect of these changes. In particular, analyses could have been conducted both for the current, less integrated system and for a single optimized system. The former would have provided an upper bound for costs, assuming continuation of the status quo. The latter would provide a lower bound for costs, which could be achieved if the existing constraints on cost minimization were eliminated.

In part, the problems with the GCES power resources studies stem from the lack of a continuous open planning process that is accessible to the public. By contrast a very different set of procedures is in place for addressing electric power issues in the Pacific Northwest concerning the Columbia River system. Both the Colorado and Columbia River systems feature extensive hydroelectric facilities operated by federal agencies that sell power to government-owned utilities under environmental constraints. But due in part to the Northwest Power Planning Act, the Northwest region has long had a major planning effort that has developed the necessary tools and institutions required to evaluate the effects of various alternatives. There is a high level of expertise on the part of federal agencies, the utilities, and other interested parties such as environmental groups and state governments. Furthermore, this process has been steadily building in expertise over the past two decades.

In contrast, the GCES and Glen Canyon EIS were the first time that much of this type of analysis had been undertaken for the Colorado River system area. In addition, given the many disparate interests, there were many procedural issues to resolve, and it has been difficult to obtain the requisite economic and financial data. Overall, the process has been difficult, time consuming, and costly. Also, unlike the process for the Columbia River, which involves a continuing mandate and an established institutional environment, it is unclear to what extent power studies for Glen Canyon Dam will continue in the future. Clearly, there is a need to update the power studies over time for the purposes of adaptive management. Moreover, without such projections, it will be difficult to plan and operate the regional power system effectively.

The BOR or Department of the Interior should sponsor the development of analytical and modeling capabilities that can continue to provide information concerning the cost of dam operations. This would permit regular revisions to reflect the rapidly evolving electricity industry and other factors. Future studies relating to the operation of Glen Canyon (and other hydro facilities) should explicitly consider how current practices and constraint may be altered by factors such as the evolution of the utility industry. Continued modeling also could facilitate the numerous processes that are now affecting the Colorado River hydro system. Subsequent to the release of the final WAPA marketing EIS in the summer of 1995, a commitment level will be established for firm power and energy to be marketed through 2004. This process may be delayed, however, by consideration of proposals to privatize WAPA. As mandated by the Grand Canyon Protection Act, WAPA has initiated a Replacement Resources Process to study and report on methods to make up for any reductions in Glen Canyon Dam output. The U.S. Energy Policy Act of 1992 requires WAPA customers to prepare and implement Integrated Resource Plans that consider a full range of supply options, including demand-side management and renewable energy sources.

Continuing study of power economics is important, given the complex issues being examined and the need for updating. Such capabilities could be used in optimizing the timing of experimental flows. This could be a major issue. The endangered fish research included as a common element in the EIS alternatives would involve monthly release volumes similar to the seasonally adjusted steady flow alternative. It is unclear how long this research would continue, but the EIS indicates that it could be for as much as 10 years.

More generally, an attractive strategy would be to experiment with highly restrictive flow regimes (e.g., seasonally adjusted steady flow) in the short term when surplus capacity is available and the cost of such alternatives is low. Then, if the costs of altered flows rise in the future, decisions on whether to move toward a less restrictive alternative (e.g., modified low fluctuating flows) could be made based on revised studies concerning the effects of flow regimes on power and other resources.

Near-term experimentation with highly restrictive flow regimes may also reduce the need to experiment in later years, when less surplus capacity is available and costs could be much higher. Thus, a strategy of extensive short-term experimentation could reduce long-term electricity costs. When evaluating future dam operations, especially experimentation with highly restrictive flow regimes, decision makers should consider long-term, as well as short-term, impacts. Experimentation which has significant long-term benefits should not be unduly restricted in an attempt to minimize short-term increases in power costs.

REFERENCES

Barrett, C. 1992a. Letter to the Editor, The Washington Post, January 21, from Clifford Barrett, Executive Director, Colorado River Energy Distributors Association.

Barrett, C. 1992b. Letter to Sheila David, Program Officer, NRC Committee to Review the Glen Canyon Environmental Studies, February 10, from Clifford Barrett, Executive Director, Colorado River Energy Distributors Association.

Bureau of Reclamation et al. April 1990. Final Report, Evaluation of Methods of Estimated Power System Impacts of Potential Changes in Glen Canyon Powerplant Operations. Bureau of Reclamation, Washington, D.C.

Bureau of Reclamation. 1991. Glen Canyon Dam, Interim Operating Criteria, Draft Environmental Assessment. Bureau of Reclamation, Upper Colorado River Regional Office, Salt Lake City.

Bureau of Reclamation. 1995. Operation of Glen Canyon Dam, Colorado River Storage Project, Final Environmental Impact Statement. Bureau of Reclamation, Salt Lake City.

Environmental Defense Fund. 1991. Estimates of Power System Impacts of Proposed Interim Flow Release Patterns at Glen Canyon Dam. Oakland, Calif.: Environmental Defense Fund.

Flaim, S.J., R.E. Howitt, and B.K. Edwards. 1994. Impacts on Irrtigated Agriculture of Changes in Electricity Costs Resulting from Western Area Power Administration's Power Marketing Alternatives. Draft report, Argonne National Laboratory, Technical Report No. W-31-109 Eng-38, for the U.S. Department of Energy, Argonne, Ill.

National Research Council. 1991. Colorado River Ecology and Dam Management. Washington, D.C.: National Academy Press.

National Research Council. 1994. Review of the Draft Environmental Impact Statement on Operation of Glen Canyon Dam. Washington, D.C.: National Academy Press.

Power Resources Committee. 1992. Power Systems Impacts of Potential Changes in Glen Canyon Powerplant Operations. Glen Canyon Environmental Studies Technical Report. Draft, Stone and Webster Management Consultants, Inc., Englewood, Colo.

Power Resources Committee. 1993. Power Systems Impacts of Potential Changes in Glen Canyon Powerplant Operations. Glen Canyon Environmental Studies Technical Report, Stone and Webster Management Consultants, Inc., Englewood, Colo.

Power Resources Committee. 1995. Power Systems Impacts of Potential Changes in Glen Canyon Powerplant Operations, Glen Canyon Environmental Studies Technical Report. Stone and Webster Consultants, Inc., Englewood Colo.

Roluti, M.J. 1993. Letter to William M. Lewis, Chair, NRC Committee to Review Glen Canyon Environmental Studies, Subject: Power System Impact of Potential Changes in Glen Canyon Environmental Studies, April 29, from Michael J. Roluti, Chair, Power Resources Committee, Bureau of Reclamation, Upper Colorado Regional Office.

Water Resources Council. 1983. Economic and Environmental Principles and Guidelines for Water and Related Land Resources Implementation Studies. Washington, D.C.: Government Printing Office.

Western Area Power Administration. 1994. Salt Lake City Area Integrated Projects Electric Power Marketing. Draft Environmental Impact Statement, Western Area Power Administration, Salt Lake City.

10

Institutional Influences on the Glen Canyon Environmental Studies

INSTITUTIONAL STRUCTURE OF GCES

Institutional frameworks influence the quality, quantity, efficiency, and cost of scientific activities. Research almost always reflects the nonscientific influences of administration, politics, bureaucracies, and, of course, funding. The Glen Canyon Environmental Studies (GCES) were strongly affected by the institutional environment within which they developed and operated. The purpose of this chapter is to explore the connection between institutional arrangements and the scientific research conducted in GCES. Factors to be considered include the structure of GCES itself, interagency conflicts and goal substitution, external oversight, and the funding and timing of research. The chapter concludes with some generalizations about GCES that might be useful lessons for other similar government research initiatives.

The internal structure of GCES changed as the organization gained experience and grew (see Chapter 2). In theory, its final configuration (after 1993) reflected a two-part management team administering multiple research groups and contractors. The GCES program manager directed the basic operations of GCES, including personnel and budget, and was the major liaison between GCES and outside agencies. The program manager conducted most of the organizational activities of GCES and coordinated the agency's interaction with the public. The second member of the management team was the senior scientist, who was responsible for direct oversight of the scientific research, including planning and execution. The senior scientist's most critical responsibility was the maintenance of scientific quality. The National Research Council (NRC) review committee, after their review of

GCES Phase I (NRC, 1987), argued that division of the administrative duties of GCES from the scientific duties would be advantageous. In practice, this division of labor turned out to be imperfect because the program manager had primary control over most aspects of the research, while the senior scientist served more as an internal critic and organizer of the intellectual effort. The senior scientist devoted 40 percent of his time to these responsibilities by mutual agreement with the Bureau of Reclamation (BOR), but experience showed that his responsibilities were more commensurate with a full-time position.

GCES was administered by the BOR, an agency of the U.S. Department of the Interior (Figures 10.1 and 10.2). Within the BOR, GCES was under the Upper Colorado Regional Office in Salt Lake City, Utah. Although as a scientific research unit, GCES was unique within the bureau, its institutional position in the region was similar to that of regional operational units such as planning, construction, and maintenance. In reporting to the regional director, the administrator of GCES was forced to deal with a portion of the BOR's hierarchy especially sensitive to local interests. As a result, the GCES administrator competed for influence and control with local water and power users who had long-established lines of communication with the office of the regional director.

Decisions about GCES often took a strongly regional perspective. For example, when contemplating scientific experimentation with releases from Glen Canyon Dam, regional administrators naturally were most concerned with the effects of operational changes on regional power marketing (WAPA, 1990). If decisions had been made at a higher level within the BOR, other considerations of broader significance could have come into play more strongly, such as the national significance of the Grand Canyon and the importance of GCES as a prototype research effort that might be necessary in other locations for other large dams operated by the bureau. Much to the BOR's credit, after considerable debate, the regional administrators agreed to forego some power revenues in order to conduct the experimental releases that were part of GCES (Patten, 1991).

GCES was part of just one of several regional offices. This arrangement became a problem when the research interests of the BOR were different from the research interests of other federal agencies within the Interior Department. Because the conduct of scientific research begins with the formulation of research issues or questions, the position of GCES determined an early focus on issues primarily of interest to the BOR. Despite the fact that other Interior agencies, particularly the National Park Service and the Fish and

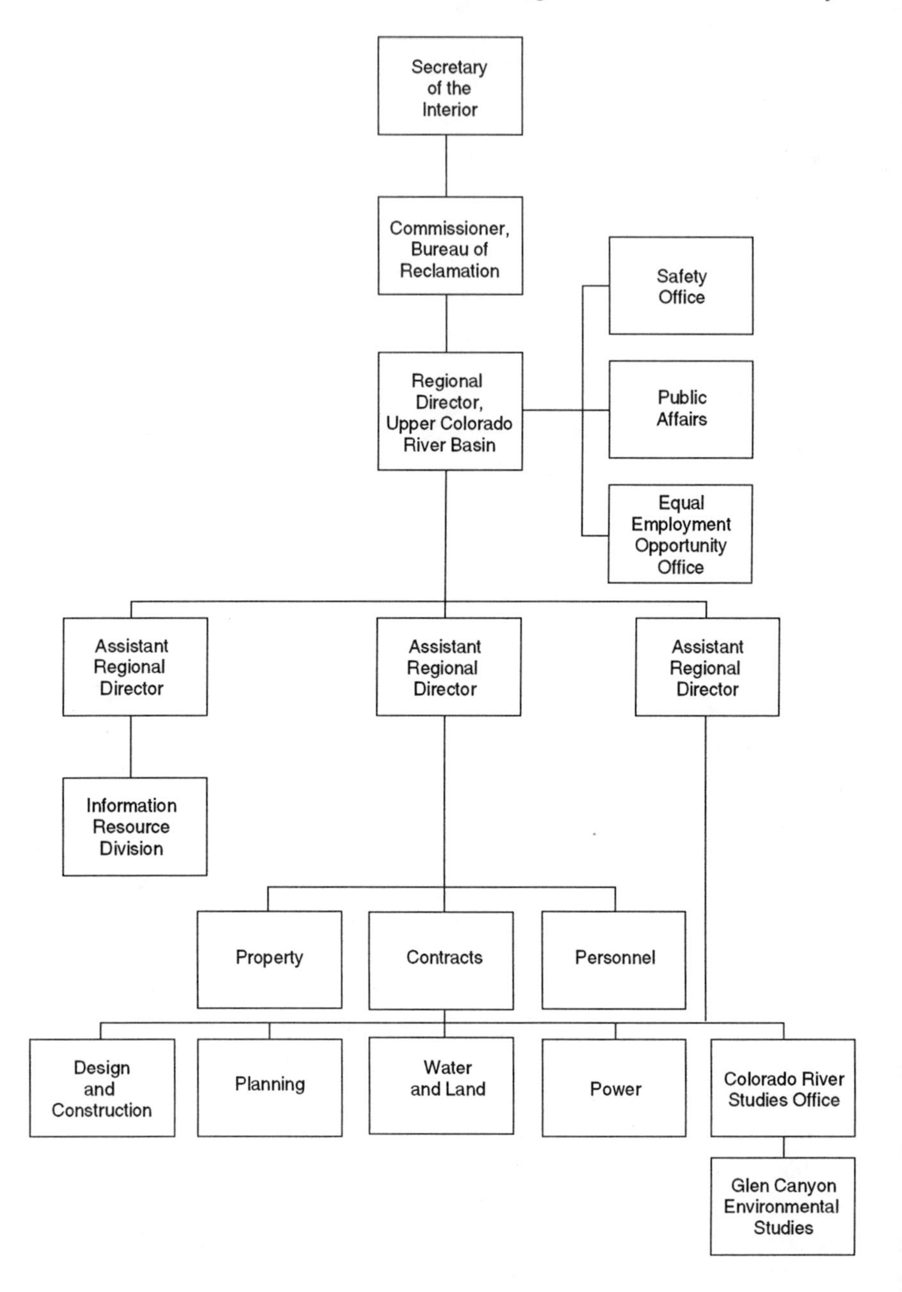

FIGURE 10.1 General organizational chart showing the bureaucratic position of GCES in the BOR and Department of the Interior. SOURCE: Redrawn from data provided by D. Wegner, Bureau of Reclamation.

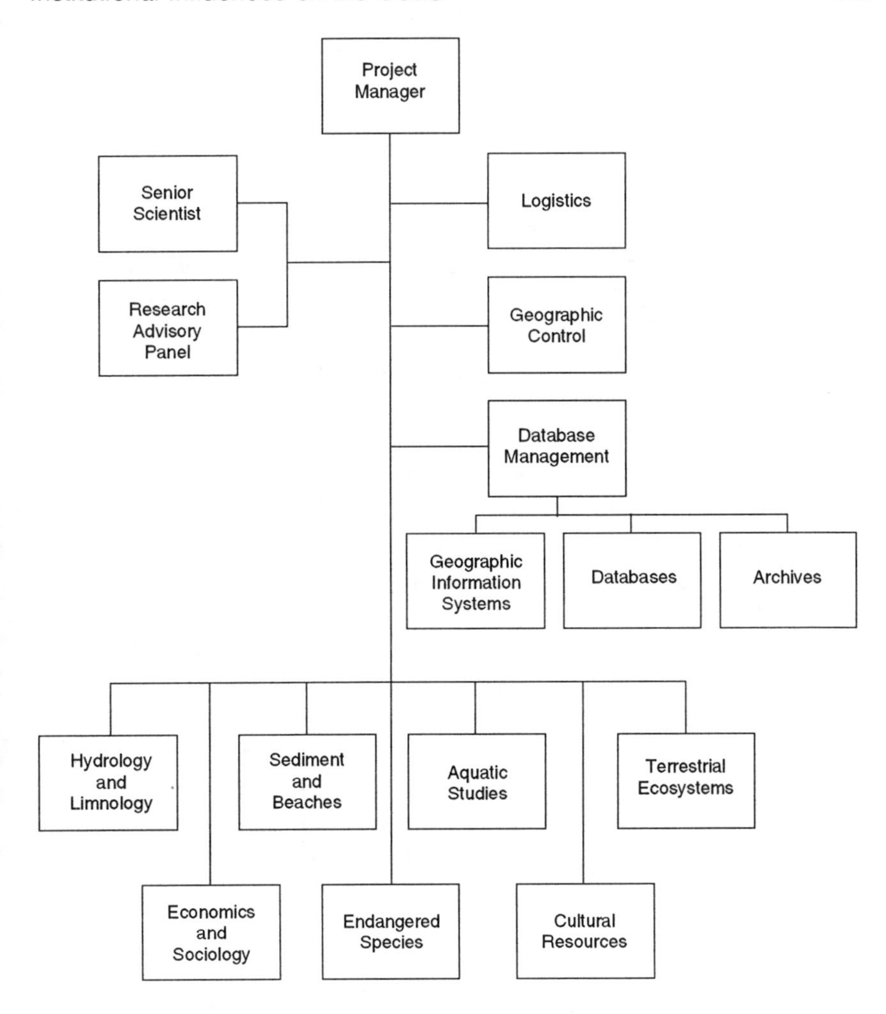

FIGURE 10.2 Organizational chart showing the general design of the GCES. SOURCE: Redrawn from data provided by D. Wegner, Bureau of Reclamation.

Wildlife Service, were included in the research, their interests were initially secondary. The original narrow focus changed over the life of GCES, with a continual widening of research activities to more directly account for the interests of other agencies (see Chapter 2). A better arrangement would have been to place GCES administratively in such a way that the research unit could fully accommodate the needs of several agencies from the beginning (NRC, 1987). The placement of the research unit in the Interior Department such that it reported to an assistant secretary would likely have resulted in better coordination of scientific research questions and a more rapid route to consensus among the various agencies. Instead, the history of GCES is marked with interagency conflicts that could have been minimized or avoided.

Placement of GCES above the regional level would also have broadened the perspective of the project. GCES developed as an effort specifically focused on Glen Canyon Dam and Grand Canyon. Researchers in other areas, however, were simultaneously investigating similar problems on other rivers. An exchange of information and ideas between these studies never took place, even though in some cases the work was being conducted under the direction of the BOR. Bureau-sponsored research on Trinity River below Trinity Dam in Northern California, for example, included specific investigations of dam operations designed to move sediment through the downstream system, and experimental flows were used in a test case very similar to those of GCES (U.S. Senate, 1984; Kondolf and Wolman, 1993). Although the GCES project manager visited the Trinity River and held conversations with workers there, no comparative studies or formal transfer of results occurred. It also appears that BOR-sponsored research on the Gunnison River was completely ignored by GCES investigators, even though the work on the Gunnison included experimental releases intended to move boulders in downstream rapids (Chase, 1992; Auble et al., 1991). While the Trinity and Gunnison rivers and their canyons are smaller landscape features than Grand Canyon, the strong similarities in research questions and the use of experimental flows argue for substantial interchange of information and ideas.

INTERAGENCY CONFLICT IN THE EVOLUTION OF GCES

During the 13 years of GCES research, conflicts developed between the BOR and the National Park Service (NPS), the Fish and Wildlife Service, various state agencies, Western Area Power Administration (WAPA), and the Colorado River Energy Distributors Association. Conflicts with NPS were probably inevitable because of the overlapping jurisdictions of the Park Service and the BOR (Johnson and Carothers, 1987). During the early phases of GCES, there was a notable lack of cooperation between the two agencies (NRC, 1987). During the later phases of GCES, cooperation improved, although research activities appeared to be separate. For example, the NPS sponsored extensive investigations of beach erosion processes, including collapse of beach faces, the erosive role of pore pressure from ground water within the beaches, and the general adjustments by beaches to changes in stream flow (Cluer, 1991). Meanwhile, research by the U.S. Geological Survey (USGS) emphasized sand transport and storage at a larger scale, as well as depositional processes that created beaches and affected archeological sites (Hereford et al., 1991). The draft final reports of all these activities do not show any significant integration of the research activities, almost as though each agency pursued its own activities without reference to the other.

Institutional barriers between the NPS and the BOR and other agencies may also have been responsible for awkward funding arrangements for individual researchers. Because the BOR contracted much of GCES research to other agencies, the most important investigators often were not BOR employees or independent contractors. Much of the archeological research, for example, was conducted by NPS employees. Rather, they owed their institutional allegiance to another agency: a most important general ecologist was an NPS employee, a most critical native fish specialist was an Arizona Department of Game and Fish (ADGF) employee, and the primary sediment transport experts were USGS employees. In all of these cases the researchers had primary responsibilities to their home agencies rather than the BOR, so when funding or scheduling conflicts arose, the bureau's position was secondary. The result was instability for GCES because important reports were delayed, and their results could not be used for midcourse corrections or further planning for other related projects.

Significant conflict developed between GCES and the Fish and Wildlife Service during the last phases of the research. This conflict focused on endangered native fish species and demonstrated the problems inherent in positioning the primary research organization low in the organizational hierarchy of the Interior Department. GCES was charged with investigating the effects of Glen Canyon Dam operations on native fishes, particularly the humpback chub. Through several years early in GCES, the BOR's contractors collected data about the fishes and began formulating conclusions (e.g., Kubly, 1990). When the Secretary of Interior decided that an environmental impact statement (EIS) was to be written concerning the dam's operations, the GCES data and conclusions were an obvious source of scientific information. The Endangered Species Act, however, required that the Fish and Wildlife Service provide an opinion concerning endangered species in the canyon (Behnke and Benson, 1980). The Fish and Wildlife Service then undertook an expensive research effort, funded through the

GCES budget, in order to develop its own conclusions.

Later, when the BOR endorsed a preferred alternative for operating the dam as part of the EIS, it relied on the conclusions of the GCES. The Fish and Wildlife Service, relying on its own research, settled on a different set of operating rules. The result was that two agencies in the Interior Department took different positions on how best to protect the endangered fish, and both positions were based on research funded through GCES. By late 1993 the issue was settled by an arrangement whereby the Fish and Wildlife Service agreed to support the BOR's position in return for a guarantee of continued funding to the NPS for investigations of the endangered-fish population. The entire effort would have been more effective if the institutional arrangements for the research had been centralized within the Department of Interior (perhaps in such an agency as the National Biological Survey), so that the research could have been focused, nonduplicative, and productive of a single defensible conclusion. Differences of opinion and interpretation in research efforts are inevitable and generally healthy, but their early resolution saves time and money.

Earlier in GCES, another conflict had developed around research on the effects of dam operations on fishes. The trout fishery immediately downstream from the dam was viewed by the state of Arizona as an important recreational benefit of the structure (Molles, 1980), and research by the ADGF had been funded by both the state and to a lesser degree by the bureau through GCES. ADGF had developed considerable scientific expertise regarding the trout and game species in Lake Powell, but when the Secretary of Interior directed the BOR to write an EIS for the dam operations, the bureau initially refused to include the state agencies as cooperators in the effort. At a symposium held in Santa Fe in 1990 on the state of knowledge for the Grand Canyon environment, the NRC committee was especially critical of the BOR's exclusive policy regarding cooperators on the environmental impact statement. Eventually, the bureau included as cooperators all the interested parties in an effort to build a consensus for a preferred alternative.

During the subsequent 3 years, in an attempt to broaden its base of support, the BOR gradually expanded its group of cooperators, and the state wildlife agencies were included. Eventually, the BOR completely reversed its exclusionary position and invited a wide range of interests to become cooperators, many of whom had direct interests in the scientific research in the canyon. This expansive policy was successful in improving the exchange of information and aided in the process of building a consensus position for a preferred alternative for operating Glen Canyon Dam. The expanded range of interests also impacted the research structure of GCES by broadening the range of questions that were asked. Native American concerns and interest in non-use values expressed by representatives in the discussions, for example, resulted in new research questions that became part of the expanded GCES agenda.

The WAPA, as the federal agency marketing hydroelectric power from Glen Canyon Dam, and the BOR, as the agency using hydroelectric power revenues from the dam to repay costs, have had a direct interest in scientific research in the canyon because the conduct and outcomes of the research might affect future dam operations (WAPA, 1990). The Colorado River Energy Distributors Association (CREDA) represents the interests of electrical consumers of a significant portion of the power generated by Glen Canyon Dam. Most of CREDA's members are rural electrical cooperatives and small towns; these consumers are highly sensitive to changes in electric rates. Hydroelectric marketers and consumers therefore were direct players in the administration of science in the canyon throughout the history of GCES. Initially, the hydroelectric interests wanted the research to be concluded as quickly as possible so that the dam could be operated to its maximum potential for electric power production (Barrett, 1992). Their position, simply stated, was that the scientific research was prohibitively expensive. They consistently made this position known to the regional director of the BOR, who oversaw the GCES, and they were always present to state their position at public meetings of the researchers and even at meetings of the NRC committee. Initially, their position was that a delay in implementing new operating rules for the dam was costly in terms of foregone power revenues. Despite the problems inherent in the interaction between the NRC committee and the GCES while research was on-going, the continuous involvement of the committee as an external, unbiased review body resulted in an improved research effort and a more fruitful expenditure of public resources. To have waited until the completion of the research and then offer committee guidance would have diminished the potential contribution of the NRC.

By 1990 two points had become clear: first, GCES research in the canyon would not be completed quickly, and, second, researchers were beginning to make the case that the dam should be operated for a time in an experimental mode that might further restrict operations for power production. At the 1990 meeting of researchers and administrators in Santa Fe and in subsequent statements, WAPA and especially CREDA began to state support publicly for scientific research in the canyon. The hydroelectric power interests obviously wanted to see reasonable answers to their questions about the effects of dam operations as a way of preventing further, long-term delays on normal operations. Short-term costs appeared to be a reasonable investment in a more stable and predictable future unencumbered by additional research.

WAPA and CREDA's members were naturally opposed to adjusting dam operations for research purposes, even in the short term, because researchers wanted flows that reduced the value of Glen Canyon Dam electrical output. Through a year-long series of negotiations by GCES managers, the regional BOR office, and WAPA, a plan for research flows was agreed on. WAPA, in discussions with the NRC committee, suggested that costs in terms of lost power revenue would exceed $30 million. Estimates varied from time to time, but the WAPA model (WAPA, 1989) was not especially accurate (Hughes, 1991; see Chapter 9). Conversations between BOR representatives and the NRC committee after the flows were complete suggested the actual foregone revenue was much less (about $3 million). During the period of research flows, CREDA raised their electrical rates by almost 40 percent and declared that a significant portion of the increase was caused by scientific research in the canyon (Barrett, 1992). The NRC committee concluded that the increased costs resulted from two other sources more important than GCES: low runoff, which resulted in reduced power production from the dam irrespective of the research flows (necessitating the purchase of more expensive replacement power), and adjustments in rates to reflect generally increasing costs that would have occurred in any event.

The lesson to be learned from the conflicts between GCES and the hydroelectric power interests is that not only is science expensive, but its costs can impinge on the interests of particular groups rather than a general unidentified population. In attempting to satisfy hydropower users, scientific researchers found it necessary to modify their plans in order to reach a compromise between what was scientifically optimal and what was politically acceptable.

GOAL SUBSTITUTION BY AGENCIES WORKING FOR GCES

Interagency conflict was not always obvious and direct during the conduct of GCES. Some agencies worked on the scientific research in the Grand Canyon as contractors for the BOR and GCES. GCES administrators assumed that the contractors would adopt GCES goals, but in some cases the NRC committee found that contracting agencies in fact were more interested in pursuing their own goals by using GCES funds as a means of support. The contracting arrangements therefore had a strong influence on the scientific products that ultimately resulted (see Chapter 2).

For example, the USGS sought to pursue its own research and monitoring interests through GCES. Stream gauges in the canyon might logically be operated as part of the national network of stream gauges on the nation's most important rivers, receiving funding from a national appropriation to the USGS for such efforts. In order to obtain critical gauge data, however, the GCES budget shouldered the cost for maintaining gauges on the Colorado River in the Grand Canyon. Thus, instead of GCES simply supporting additional needed research on using and interpreting the data, it paid for the initial collection. In some cases the USGS had research interests in or near the canyon, and particular researchers desired support for special projects they wanted funded. Unable to support research into historical photographic sites, extended investigations into debris flow processes, exploratory flow models, and some tributary studies, the USGS sought funding for these from GCES. These topics were potentially of interest to GCES but in some cases did not have high priority. Instead of working directly with GCES to develop a coordinated series of projects specifically targeted to BOR's needs, the USGS proposed unrelated projects reflecting its own interests. Eventually, GCES declined to fund some proposed projects or funded others only briefly, but the end result was a poorly integrated research effort in the earth and water science areas.

Agency ties to GCES were not wholly disadvantageous. In the final analysis, some of the best science in the GCES program was that derived from the work of the USGS, but the relationship of GCES to USGS was an uneasy one.

An example of successful communication between agencies also involves the USGS. After the first phase of GCES, it became apparent that BOR's models of the dynamics of sediment in the canyon failed to describe observed conditions and that they were not useful for predictive purposes (NRC, 1987; see Chapter 5). After discussions with GCES personnel, USGS researchers established specific projects that would show how much sediment was moving through the system, how it was deposited in pools along the canyon, and how it was moved to beaches during high flows (Schmidt and Graf, 1990). Using observations of actual processes in the river rather than abstract modeling, the USGS effort not only successfully contributed to basic scientific knowledge about river processes but also contributed useful applied knowledge that the BOR later used in considering options for dam operations (BOR, 1993). It was an example of science at its best because the goals of the funding agency (the BOR) and the research group (the USGS) were similar and had been agreed on prior to the research.

Goal substitution also occurred within the BOR when the need for an EIS was announced. As the best source of information about the Glen Canyon Dam and the Colorado River, GCES managers and researchers immediately became involved in the preparation of the EIS. GCES administrators began to manage research that supported the needs of the EIS, particularly as related to cultural and archeological resources. A great deal of GCES intellectual effort went into debating the appropriate flow option that would be designated as the preferred alternative in the EIS.

As a result of these institutional arrangements, the goals of the BOR's EIS supplanted the goals of GCES, and the conduct of science was diverted from the long-term perspectives of GCES to the short-term perspectives of the EIS. Consistent preoccupation with short-term goals to the detriment of useful long-term research has been common in BOR research (Leopold, 1991). One casualty of the emphasis on the short term may have been the long-term monitoring plan (Patten, 1993), which was originally conceived as a precise product of GCES research. With the advent of the EIS, however, the long-term monitoring plan became part of the EIS. The plan was highly general rather than specific (NRC, 1994).

One outcome of the fragmented GCES research spread throughout several agencies was a remarkable lack of integration of results. Because contracting agencies partly followed their own agendas and sometimes their own time tables (private contractors were more responsive to GCES requirements), GCES Phase II ended without a final integrated report. This shortcoming was especially serious because the basic philosophy of the studies was that they were ecosystem studies that not only provided understanding of the various components of the natural and artificial environment of the canyon but that they also explored the connections among those components. Part of the lack of integration, however, was also due to inadequate planning by GCES management. Integration must be a part of the effort from the beginning, rather than viewed as only the final task. Interim progress reports can be used during the research itself to begin the early development of an integrated perspective that can grow and mature as the project progresses.

EXTERNAL REVIEW—GCES AND THE NRC

The assurance of quality in scientific research always requires external oversight and review. Research is most credible if it bears the scrutiny of independent reviewers who can detect errors of fact or judgment, improve the final product with constructive criticism, and prevent poorly executed work from being released. The peer review system for scientific publication is a means for accomplishing such quality control. Agencies within the federal government that produce research sometimes (but not always; USGS, 1991) attempt to control quality with exclusively internal review systems, but this practice is detrimental to research in the long run.

When the BOR began the GCES work in 1982, no external oversight or review was established, but in 1986 the bureau contracted with the NRC to provide scientific advice and to evaluate the written products that had resulted from research during the first phase of the effort. This review effort by the NRC was partly the result of a need for credibility as the BOR presented its conclusions to the public and to other components of the federal and state governments. When the NRC's Committee to Review the Glen Canyon Environmental Studies gave GCES research products decidedly mixed reviews (NRC, 1987), the BOR began a new round of research and extended its relationship with the NRC. The committee expanded its oversight and advice to include review of planning efforts by GCES for its second phase and review of the draft EIS. A unique relationship developed between GCES and the NRC committee because the committee, unlike the usual scientific reviewer who sees only the end product, became directly involved in all aspects of the research—planning, execution, and reporting of the final results.

The NRC committee met periodically with GCES managers and researchers, read planning documents, evaluated specific research proposals, and reviewed written drafts of research reports. The committee provided comments, advice, evaluations through discussions, letter reports, and formal published reports.

The advice of the NRC committee was not always accepted. A specific example of the lack of meaningful response to review comments concerned the BOR's power resources studies. When the NRC committee made extensive comments on the draft report by the BOR researchers (NRC, 1992), the researchers' response was primarily to defend what they had done and not make corrective adjustments (Roluti, 1993; Chapter 9). Thus, rather than working in a give-and-take environment in which both parties were pursuing the highest-quality product, it seemed that the two participants were sometimes in an adversarial relationship. Such adversarial relationships between an NRC committee and its client agency are not unusual, but in most other cases the tension has decreased as the work progressed (Graf, 1993).

One problem with the relationship between the NRC committee and GCES was the dual role of the committee in providing advice during the research and in providing judgments at the conclusion of the work. When the committee made recommendations that subsequently were not followed, the committee was highly critical. For example, the committee strongly urged GCES to take into account nonuse values in its calculations regarding the economics of power generation by Glen Canyon Dam. GCES was slow to include such approaches in its work, and this engendered sharply negative responses from the committee. Eventually the BOR included nonuse values in its economic studies (Colby and Goodman, 1993), but the results lost some of their effectiveness because they came very late.

The NRC committee review process is not especially well suited to providing advice for research in progress, because all NRC reports, including brief letter reports, must pass through an extensive review and evaluation system of their own. The process is reasonably efficient when compared to similar arrangements in other organizations, but it still requires about 2 months. As a result, the advice needed by a sponsoring agency may be stale by the time it arrives, particularly if the work is seasonal. The hydrological and ecological systems that concerned GCES imposed many constraints on timing, which were also complicated by administrative time tables and the complicated process of meshing release schedules for the dam, research needs, and the EIS schedule (BOR, 1991). While research is in progress, the comments of a review body can sometimes be more disruptive than helpful. In any case, midcourse corrections in research depend on timely submission of documents for review and expeditious handling of the documents by the review agency.

There were notable successes in the relationship between GCES and the NRC committee. GCES managers and researchers sometimes used the committee as a sounding board for ideas, and the intellectual exchanges often were of high quality. Presentations of research and results in oral form provided workers with an opportunity to refine their thinking before going on to other more public forums. Occasional contact with GCES managers and researchers provided the committee members with insights that allowed them to function much more efficiently in their evaluations than otherwise would have been possible. Involvement of the committee in early planning for the second phase of GCES studies resulted in several substantial improvements that carried through the remainder of the project, including the establishment of the Office of the Senior Scientist and the effort to diversify the research contracting process. The committee played a key role in, for example, pointing out critical and unacceptable weaknesses in the original studies of sediment transport, identifying the need for a senior scientist on the management team, indicating the significance of omissions of nonuse values in the power economics studies, endorsing research flows, broadening the geographic scope of the work, and calling for external review of the research by advisers other than the senior scientist.

THE ROLE OF FUNDING IN GCES

During the life of GCES, there were always two sets of plans: research and finance. The financial planning was, to a remarkable degree, unpredictable on an annual basis and outside the control of GCES managers. Annual fluctuations were considerable (Figure 10.3). Because the exact amount of support expected for the following year was often unknown, the conduct of multiyear research efforts was a risky business. The long-term requirements of natural science research and the short-term planning for agency budgets often conflicted with each other. For example, assessments of chemical characteristics of water, sediment, and biological samples require a multiyear effort. Selection of sample sites, initial collection of samples, preparation of materials, and ancillary measurements precede the laboratory chemical analysis. Frequently, repeat sampling is required to obtain an understanding of the chemical stability of the system involved. When GCES workers assessed chemical contents of biological materials, they encountered problems in an important part of the work because the length of time needed for the research was longer than the annual funding cycle. When funding for the USGS, the agency in charge of the chemical analysis, decreased during GCES, the samples were available, but there was no money to analyze. Research on native fishes was especially constrained by the short-term nature of predictable funding levels.

The annual funding cycle for the research was an outcome of the institutional arrangements for the work. GCES received its funding from power revenues generated by the operation of Glen Canyon Dam through the regional office of the BOR in Salt Lake City. The level of funding made available to GCES was therefore a function of the internal priorities of the re-

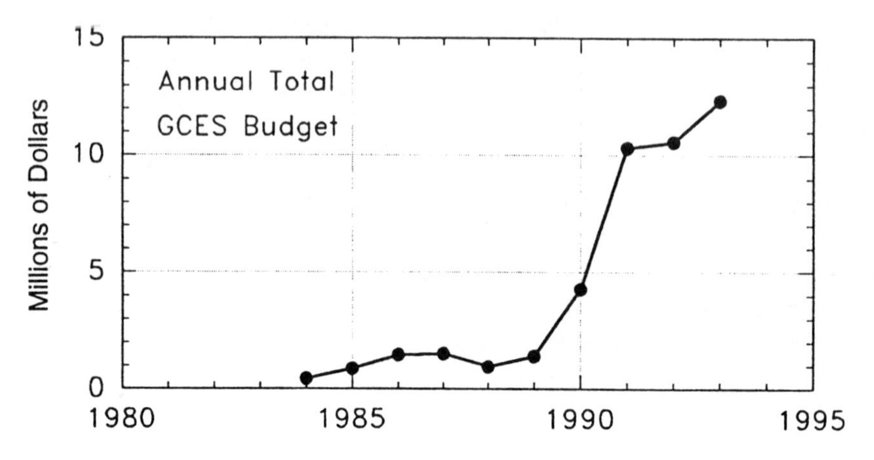

FIGURE 10.3 Annual funding history of GCES. SOURCE: From data provided by D. Wegner, Bureau of Reclamation.

gional office and of the revenues generated from a source that was somewhat variable from one year to the next because of variations in water flows and market conditions. WAPA (1988) predictions of power revenues were made by using questionable assumptions (Hughes, 1991), which further complicated the financial picture.

The mismatch between short-term budgets and long-term research might have been rectified by a multiyear funding scheme similar to the approach used by the National Science Foundation, the National Aeronautics and Space Administration, the Agricultural Research Service, and other federal agencies. These organizations authorize research for a particular level of funding over several years. Each year, as the federal budget is approved, the agency funds the next installment in the grant or contract. While GCES attempted to replicate this approach, funding uncertainties made short-term financial support the order of the day, to the detriment of more appropriate and more stable multiyear commitments (Leopold, 1991).

Legal requirements that directed certain funds for particular purposes also constrained the management of GCES research. During GCES Phase I, most of the research was determined by the need to understand the Grand Canyon environment. Funding for Phase II research was much higher than for the earlier work (Figure 10.3), but the total budget is somewhat misleading. Beginning in 1989, increasing amounts of the total GCES budget were apportioned to mandated research consisting of investigations required by law or policy (Figure 10.4; see Chapter 2). Occasionally, mandated research produced information not directly useful to GCES or produced monopolistic research rights for other agencies to deal with issues and data within the legitimate scope of GCES. When numerous archeological sites appeared to be endangered by dam operations in the canyon, laws related to the preservation of antiquities came into play that demanded some expenditure of funds to assess the sites. When the Department of the Interior decided to produce an EIS and use the GCES framework for the funding of supporting research, the diversion of funds from standard scientific research to mandated research became even more pronounced, especially with regard to endangered native fishes.

At first archaeological studies were conducted by the National Park Service (NPS) pursuant to various federal laws and did not include participation by Native American Tribes. Then, various Indian tribes with interests and history in the Grand Canyon received financial support to conduct investigations of cultural connections to sites in the canyon that were of religious or historical significance. Funding for such work was from the GCES budget, which was augmented for the purpose, but administrative costs for the work, particularly time, effort, and coordination, were extensive. By policy, the federal government preferred to offer tribes the opportunity to train their own members to conduct as much of the work as possible. Some tribes preferred to contract at least part of their investigative work to researchers outside the tribe, but the end result was that by 1992 less than half of the total GCES budget was allotted to scientific research outside the mandates superimposed on the original goals of GCES (Figure 10.4). Whether or not this research caused a decline in the quality of other science in the GCES effort is unclear.

A funding issue that plagued GCES from start to finish was the manner in which contracts were arranged. During the first phase of the research, the BOR contracted almost exclusively with its own investigators or with government agencies with which it had close relationships. The pool of potential researchers for any given part of the project was therefore limited, and reviewers of the early research questioned the failure of the BOR to more widely advertise for bids on the proposed work (NRC, 1987). One objective of the senior scientist was to open the process of requesting proposals to a broadly defined research community that included government workers, private companies, university researchers, and individuals. During the second phase of research, however, the range of investigators was only moderately more general than in the first phase. The contracting and adver-

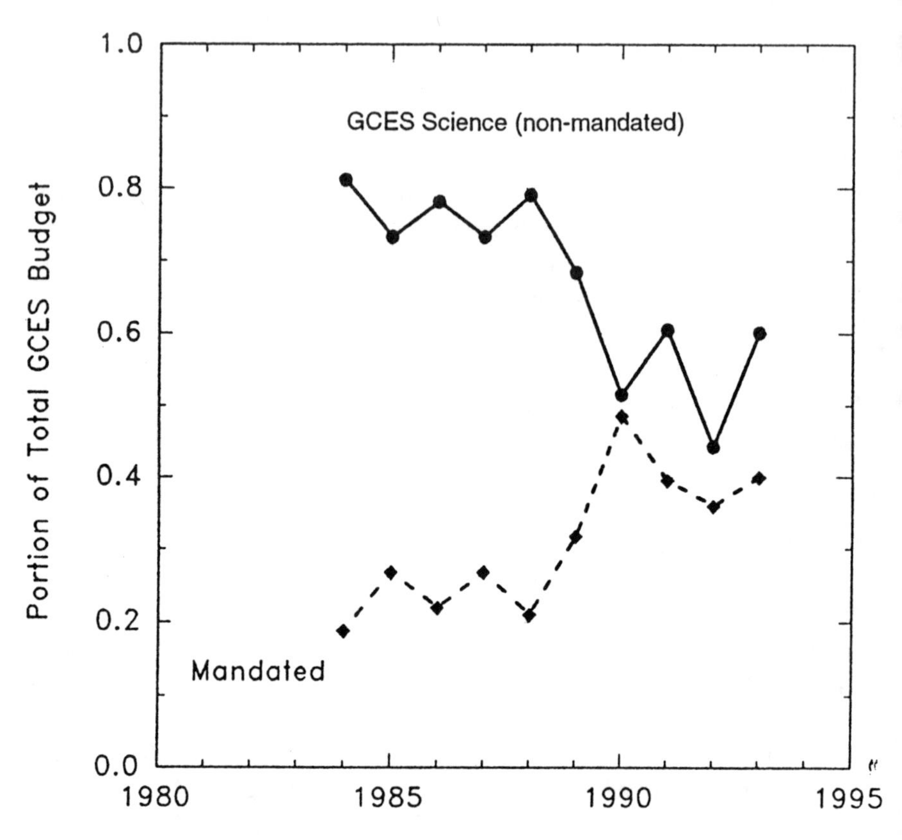

FIGURE 10.4 Portion of GCES annual budget allotted directly to GCES science and to mandated research required by other laws or customs. SOURCE: From data provided by D. Wegner, Bureau of Reclamation.

tising requirements for federal agencies turned out to be so cumbersome that it was impossible to secure the services of researchers through broad solicitation and still meet the time limitations imposed by the annual funding process. As a result, qualified researchers and organizations not in direct contact with GCES had no opportunity to bid on the work, and the BOR had no assurance that it ultimately contracted with the most cost-efficient or scientifically effective workers. The GCES experience suggests that competitive bidding for scientific research should be pursued by federal agencies as a means of controlling costs and assuring quality.

Management of GCES also lacked control over the reporting of results by federal agencies. GCES administrators reported that they were unable to stop payments to agencies that were late in generating reports. If a contracting agency failed to produce a report, GCES could not withhold payment and had to continue payments into the next year in order to obtain data and results. Thus, the constraints and controls available in contracts with private agencies seemed not to be available for public agencies. As a result, government agencies had much less incentive to produce their work in a timely fashion and in fact may have had an incentive to delay their production in order to obtain extended funding.

TIME CONSTRAINTS IN GCES RESEARCH

Time constraints posed as many problems for GCES research as did financial considerations. Timing affected the research because of unforeseen changes in the natural system and uncertainty about the amount of time available to conduct the research. Rigid planning and inability to make midcourse changes in research severely reduced the effectiveness of the first phase of GCES because of the timing of major changes in the ecosystem resulting from the 1983 flood (NRC, 1991). While it might be argued that there was no way to predict the timing of the flood, which was actually a reservoir spill, research planning should have taken into account the possibility that drastic changes might occur during the project. Although the timing of the flood was a research opportunity, it was treated as an unwanted intrusion on the conduct of the research. GCES managers learned a great deal from the failure to plan adequately for the flood, and in subsequent work they adopted innovative plans for the research that not only allowed them to accommodate radical changes in discharge but that actually called for such changes. Researchers adjusted their own work to the timing of the natural processes and, more importantly, to those processes controlled by experimental releases of water from the dam (Glen Canyon Environmental Studies, 1992). Researchers therefore used time in the second phase much more effectively because they were more flexible in at least some aspects.

Specific time limitations for GCES research were the predictable outcome of research conducted within budgets, but throughout the history of the studies there was considerable uncertainty about the duration of the research effort. Especially during the second phase of GCES, managers of the work were uncertain each year whether the work would continue the next year.

This uncertainty forced many researchers to adopt designs focused on quick results rather than the best results. When the Department of Interior decided to generate an EIS, its established schedule added more short-term thinking to GCES, which was the primary data source for the EIS. It was only with the advent of the long-term monitoring plan, mandated by the Grand Canyon Protection Act of 1992, that planning horizons expanded to more realistic proportions.

FUTURE INSTITUTIONS

The Department of Interior is now considering the formation of a research center, based in Flagstaff, to house the administrative entity that will conduct the long-term monitoring program and research associated with Glen Canyon Dam and its operation. This new entity, anticipated to report to the Assistant Secretary for Water and Science, will inherit the data and other products of the GCES, and will be an important part of the adaptive management program because it will supply data and expertise for the interpretation of the data. The research center will have the same needs for long-term planning, sound use of scientific methods, and external review as GCES required previously. Of particular importance is the need for an external review panel of independent scientists who can offer credibility to the center's research and who can introduce new ideas. All major research foundations, museums, and experimental facilities have such panels, and the Glen Canyon research center would require one to be considered legitimate by the scientific community.

RECOMMENDATIONS

The quantity, quality, and usefulness of scientific research are partly the result of the institutions within which the work takes place. Despite a varied history, the GCES made substantial contributions to basic and applied science for the Grand Canyon environment, as shown by this report. The following recommendations may be useful for the future:

1. Organizations such as GCES should be located within the Department of Interior at an appropriate level that reports to an assistant secretary to ensure the efficient flow of funding, plans, information, and products.

2. Competitive bidding for all future research in Glen Canyon and similar areas should be open to all qualified agencies and individuals to ensure that the best and least expensive alternatives are used.

3. As a means of quality control, contracts with other government agencies should be designed to ensure transfer of funds from GCES only upon the delivery of products and reports.

4. Deadlines for completion of scientific research should be clearly specified by GCES and rigorously enforced through contract mechanisms. Scientists, whether working in government agencies or as private consultants, should all be held to the same standard: work should be completed on time and within original budget estimates. If it is not, GCES should terminate further funding and seek remedies for the deficiencies.

5. Studies such as GCES should take into account similar research being conducted in other areas.

6. Long-term planning (several years) is essential for effective research and wise use of financial support.

7. In deciding which contractors to support, projects such as GCES should evaluate not only the quality of proposals but also the degree to which the proposed work directly supports project objectives.

8. The funding of research should be based on management needs, not on perceived political requirements.

9. Final integration of projects such as GCES research should be an integral part of the research plan. The essence of the ecosystem approach to research and adaptive management is the definition of relationships and connections among the elements of the system. The GCES effort explicated individual elements without connection to others.

10. For projects of broad scope or long duration, the position of senior scientist should be full time rather than part time.

11. Any internal research center administered by the BOR for the purpose of managing the continuing Glen Canyon research (such as the proposed research center at Flagstaff) should have the benefit of an external oversight and review board to provide unbiased advice and perspective.

REFERENCES

Auble, G.T., J. Friedman, and M.L. Scott. 1991. Riparian Vegetation of the Black Canyon of the Gunnison River, Colorado: Composition and Response to Selected Hydrologic Regimes Based on a Direct Gradient Assessment Model. U.S. Fish and Wildlife Service Report for the National Park Service Water Resources Division, National Park Service, Ft. Collins, Colo.

Barrett, C. 1992. Letter to Sheila David, Program Officer, NRC Committee to Review Glen Canyon Environmental Studies, February 10, from Clifford Barrett, Executive Director, Colorado River Energy Distributors Association.

Behnke, R.J., and D.E. Benson. 1980. Endangered and Threatened Fishes of the Upper Colorado River Basin. U.S. Department of Agriculture, Cooperative Extension Service, Bulletin 503A, Colorado State University, Fort Collins, Colo.

Bureau of Reclamation. 1991. Glen Canyon Dam, Interim Operating Criteria, Draft Environmental Assessment. Bureau of Reclamation, Upper Colorado River Regional Office, Salt Lake City.

Bureau of Reclamation. 1993. Operation of Glen Canyon Dam, Colorado River Storage Project, Arizona. Draft Environmental Impact Statement, 3 vols, Bureau of Reclamation, Upper Colorado River Regional Office, Salt Lake City.

Chase, K.J. 1992. Gunnison River Thresholds for Gravel and Cobble Motion, Black Canyon of the Gunnison National Monument. M.S. thesis, Colorado State University, Ft. Collins.

Cluer, B.L. 1991. Daily Responses of Colorado River Sand Bars to Glen Canyon Dam Test Flows, Grand Canyon, Arizona. National Park Service, Grand Canyon National Park, Ariz.

Colby, B.G., and I. Goodman. 1993. Memorandum to National Research Council GCES Committee, Summary of Non-Use Values Peer Review Meeting, August 23.

Glen Canyon Environmental Studies. 1992. Glen Canyon Dam, Interim Operations, Interim Flow Monitoring Program. Glen Canyon Environmental Studies Office, Flagstaff, Ariz.

Graf, W.L. 1993. Landscapes, commodities, and ecosystems: the relationship between policy and science for American rivers. Pp. 11-42 in Sustaining our Water Resources. Washington, D.C.: National Academy Press.

Hereford, R., H. Fairley, K. Thompson, and J. Balsom. 1991. The Effect of Regulated Flows on Erosion of Archeologic Sites at Four Areas in Eastern Grand Canyon National Park, Arizona: A Preliminary Analysis. U.S. Geological Survey Report for Glen Canyon Environmental Studies, U.S. Geological Survey, Flagstaff, Ariz.

Hughes, T.C. 1991. Reservoir operations. Pp. 207-225 in Colorado River Ecology and Dam Management. Washington, D.C.: National Academy Press.

Johnson, R.R., and S.W. Carothers. 1987. External threats: the dilemma of resource management on the Colorado River in Grand Canyon National Park, USA. Environmental Management 11:99-107.

Kondolf, M., and M.G. Wolman. 1993. The sizes of salmonid spawning gravels. Water Resources Research 9(7):2275-2285.

Kubly, D.M. 1990. The Endangered Humpback Chub (Gila cypha) in Arizona: A Review of Past Studies and Suggestions for Future Research. Bureau of Reclamation, Salt Lake City.

Leopold, L.B. 1991. Closing remarks. Pp. 254-257 in Colorado River Ecology and Dam Management. Washington, D.C.: National Academy Press.

Molles, M. 1980. The impacts of habitat alterations and introduced species on the native fishes of the Upper Colorado River basin. Pp. 163-181 in Energy Development in the Southwest, vol. 2, W.D. Spofford, A.L. Parker, and A.V. Kneese, eds. Baltimore: Johns Hopkins University Press.

National Research Council. 1987. River and Dam Management: A Review of the Bureau of Reclamation's Glen Canyon Environmental Studies. Washington, D.C.: National Academy Press.

National Research Council. 1991. Colorado River Ecology and Dam Management. Washington, D.C.: National Academy Press.

National Research Council. 1992. Letter report to Michael Roluti, Bureau of Reclamation, October 21, 1992, Committee to Review the Glen Canyon Environmental Studies comments on May 1992 draft report "Power System Impacts of Potential Changes in Glen Canyon Power Plant Operations." Water Science and Technology Board, National Research Council, Washington, D.C.

National Research Council. 1994. Review of the Draft Federal Long-Term Monitoring Plan for the Colorado River Below Glen Canyon Dam. Washington, D.C.: National Academy Press.

Patten, D.T. 1991. Glen Canyon Environmental Studies research program: past, present and future. Pp. 239-253 in Colorado River Ecology and Dam Management. Washington, D.C.: National Academy Press.

Patten, D.T. 1993. Long-Term Monitoring in the Grand Canyon: Response to Operations of Glen Canyon Dam. Draft, Glen Canyon Environmental Studies, Tempe, Ariz.

Roluti, M.J. 1993. Letter to William M. Lewis, Chair, NRC Committee to Review Glen Canyon Environmental Studies, Subject: Power System Impact of Potential Changes in Glen Canyon Environmental Studies, April 29, from Michael J. Roluti, Chair, Power Resources Committee, Bureau of Reclamation, Upper Colorado Regional Office.

Schmidt, J.C., and J.B. Graf. 1990. Aggradation and Degradation of Alluvial Sand Deposits, 1965 to 1986, Colorado River, Grand Canyon National Park, Arizona. U.S. Geological Survey Professional Paper 1493, U.S. Geological Survey, Washington, D.C.

U.S. Geological Survey. 1991. Guide for Authors of Reports of the U.S. Geological Survey. U.S. Geological Survey, Washington, D.C.

U.S. Senate. 1984. Fish and Wildlife Restoration in the Trinity River. Senate Report 98-647, to Accompany H.R. 1438, Calendar No. 1295, 98th Congr., 2nd Sess.

Western Area Power Administration. 1988. Analysis of Alternative Release Rates at Glen Canyon Dam. Salt Lake City: Western Area Power Administration.

Western Area Power Administration. 1989. Analysis of Proposed Interim Releases at Glen Canyon Powerplants. Salt Lake City: Western Area Power Administration.

Western Area Power Administration. 1990. Glen Canyon Environmental Studies Research Releases: Economic Analysis. Salt Lake City: Western Area Power Administration.

11

Lessons of the Glen Canyon Environmental Studies

INTRODUCTION

Federal management of water is undergoing a maturational change that involves a drastic reduction in the number of new water projects and an increase in emphasis on qualitative aspects of water management (Wilkinson, 1993). The leadership of the Bureau of Reclamation (BOR) has acknowledged and accepted the necessity for this change, although the institutional characteristics of the bureau cannot be expected to adapt overnight to a new mission.

Qualitative aspects of water management include improvements in efficiency of water use as well as adaptation of water management to a broad range of environmental objectives such as those that are apparent from the Glen Canyon Environmental Studies (GCES). In dealing broadly with environmental issues, the BOR must find ways to work efficiently with other agencies that have primary expertise in and responsibility for specific kinds of environmental resources. Thus, the GCES has, in microcosm, been a test of the proposition that the BOR can execute a broad-ranging cooperative environmental study of a large river ecosystem and produce results that are useful to management.

Previous chapters have illustrated various weaknesses in the organization and execution of GCES. In some instances, these weaknesses may be peculiar to the circumstances of GCES. In other instances, the GCES has shown why some strategies are doomed to failure while others have a much higher chance of success. This chapter offers generalizations from the experience of GCES, in anticipation that the BOR and other government agencies must study complex environmental systems prior to developing man agement strategies that take into account diverse kinds of resources.

It is easy to focus on the defects of a complex project such as GCES. It would be a mistake, however, to overlook the milestones of achievement that the BOR passed in improving GCES and its institutional underpinnings. The achievements of BOR through GCES missions will be the focus of the last part of this chapter.

ELEMENTS OF A USEFUL ECOSYSTEM ANALYSIS

GCES has illustrated that a successful and cost-effective ecosystem analysis of use to management must meet a variety of requirements that extend well beyond the research plan, data collection, and data analysis. Management-oriented studies of environmental systems can be more difficult to organize than academic studies because they must operate within the institutional framework of mission agencies, be consistent with a variety of laws not directed to ecosystem management, reflect the interest of constituencies that affect government, be subject to strong constraints of time and budget, and produce results that are immediately useful to management. Thus, the elements of a successful study involve organizational and administrative matters as well as scientific ones.

The Planning Sequence

The planning sequence for a successful ecosystem analysis must include steps that take advantage of existing information, define the scope as tightly as possible but still realistically with respect to programmatic objectives, and project the products of analysis and the schedule on which they can be delivered. The planning sequence should begin with the creation of a planning group that is selected for its expertise in the major subject areas to be studied. The planning group should include at least one individual having expertise in each of the major areas of study, as well as several individuals who have experience in the integrative collection or interpretation of information from different areas of study. If the planning group forms primarily around vested interests or agencies rather than the needs of the project, the plan will likely be flawed.

Review of Existing Information

Planning begins with an intensive review of information not only on the location to be studied but also on other similar kinds of systems. This phase of preparation should culminate in a synopsis of all existing available information, both published and unpublished. This step consumes time (perhaps as much as a year), but it is the only means by which the collective experiences of individuals who have dealt previously with similar issues can be brought to bear on the creation of a study plan. GCES proceeded without this phase, and in many instances the issues of GCES were treated as if they were entirely novel, whereas in fact the environmental issues associated with the operation of large dams are recurrent and have been studied extensively in the western United States.

Definition of Scope

Following extensive review of existing information, the planning group will need a list of resources, a list of management options, and an ecosystem diagram (Chapter 2). These three items are the basis for the definition of scope and the study plan. As shown by GCES, most of the resources to be listed for a particular site will be obvious, but preparation of the list may present some unexpected difficulties. For GCES, the nonuse value of the Colorado River corridor below Lake Powell was not originally listed as a resource because administrative policy precluded its recognition until GCES was almost complete. In addition, cultural resources and effects on tribes were not considered until later in the GCES. Obviously, any intentional or inadvertent exclusions from the list of resources will ultimately undermine the utility of the analysis.

Preparation of the list of management options is also critical, and GCES demonstrated that its preparation can be even more difficult than the preparation of the list of resources. Where particular kinds of management options have not been studied administratively or are not favored by a sponsoring agency, they may be precluded on the grounds that even listing them or studying them would seem to legitimize their use. It is essential that this mentality be discarded if the analysis of management options is to be successful. At the same time, individuals conducting studies or using the results of studies should realize that mere consideration of management options does not necessarily justify their implementation, which may be complicated by legal, financial, or political factors that lie outside the realm of analysis. For example, one of the major factors impairing the cost efficiency of GCES was the inability of the BOR to separate the hypothetical range of management options from the range of management options preferred by or acceptable to the bureau and its cooperators.

GCES shows the peril of studies that are organized primarily around a list of resources and management options, even though such lists are integral to the formulation of the study plan. The context for resources and management options in an ecosystem analysis is the ecosystem diagram. Without a diagram, the study plan will be flawed in its failure to consider the causal connections between ecosystem components. Such connections must be understood before the outcomes of management options can be predicted.

Ecosystem diagrams can be quite simplistic, in which case they may be essentially useless. Boxes with the names of resources connected by lines showing all possible pairwise combinations are not helpful. The ecosystem diagram needs to be subjected to intensive scrutiny and debate among members of the planning group and should be reviewed by individuals (experts) outside the group who are already familiar with the resources or the system. Causal connections that are essential or critical to an understanding of the system should be distinguished from those that are not so critical. The pathways of influence for management options should be identified because they will be of particular interest in final use of the analysis for predictive purposes.

GCES adopted the ecosystem concept (Chapter 2) but did not use it effectively because it came too late and was not treated as a true driving force for the study design even after it was adopted. The ecosystem diagram is meaningless if it is used as window dressing or as justification a posteriori rather than as a planning tool.

After the list of resources and management options together with the ecosystem diagram are in place, the study group should return to a synopsis of existing information and focus on that which is already available for the site to be studied. The planning group should then decide whether or not existing information is likely to be useful in explaining some of the causal connections shown in the ecosystem diagram. This may involve consultation with specialists who collected information in the past or who are familiar with particular kinds of data analysis.

Using What Is Already Known

Newly authorized studies of environmental systems often proceed as if information collected in the past at the study site is totally irrelevant. This was the case with GCES, which was criticized by the National Research Council (NRC, 1987) for having ignored extensive past data collection on the Colorado River and Lake Powell. The NRC assisted GCES by sponsoring a workshop in 1990 to which authorities on the resources of the Grand Canyon were invited and asked to summarize the existing state of knowledge about the resources of the GCES study area. Had this been done early in the planning of GCES, it would have been far more useful to the program. The use of past information could have extended even further to the analysis of existing records on sediment, temperature, and chemical concentrations and biota of the Colorado River and Lake Powell. In the rush to begin new work, planning groups characteristically are tempted to waive a hard look at existing information, and this leads to wasted resources and unnecessary repetition of the elementary phases of ecosystem analysis from one study to another.

Implementation of the Plan

Formation of the Study Group

The planning group should give way to a study group. One fault with many complex studies is that the planning group becomes the study group. Because the planning group is selected before the dimensions of the study are known, its composition may render it ineffective as a study group. Furthermore, one criterion for inclusion as a member of a study group should be successful competition in a proposal solicitation process that is open to government employees, public-sector contractors, and universities. Thus, the study group is flexible and is dictated in any given phase of the study by the requirements of the study, not by membership in the planning group or other factors not related to successful completion of the study.

Contracting and Project Leadership

The GCES showed some of its most severe flaws in the implementation phases involving contracting and formation of the study group. Government agencies participated in the planning phase on the grounds of their vested interests as shown by their statutory responsibilities. While this is understandable and defensible, its continuation into the study phase essentially closed off flexibility in the solicitation of proposals or in the optimization of the study group to meet the needs of the study. In effect, agency missions were taken by GCES as entitlement for funding (Chapters 2 and 10). This in turn led to other problems, including the inability of project management to demand performance as contractually agreed or to redirect funds when performance of a particular contractor was deemed inadequate or of low priority.

The ideal ecosystem analysis would require a high degree of authority centralized in the project manager. The project manager for large studies might need to be assisted continuously by a senior scientist, as was recommended for GCES, simply because the management of the business component for a project of broad scope can compete with oversight of scientific dimensions of the study. The project manager can function most effectively, and with lowest cost, without obligations to provide support to any entity or individual or to continue supporting activities that prove to be inadequate or unnecessary. Such flexibility was absent in GCES, and the result was unreasonable distortion of project scope, failure of federal agencies to meet contractual obligations while continuing to receive support, and excessive focus on budget continuity rather than project objectives. GCES shows clearly that the public sector, like the private sector, cannot function efficiently unless there is a continuous element of merit-based competition in the award of support and in its continuation.

Outside Advice

Another essential element of the implementation phase is an external advisory board that is retained exclusively for the purpose of providing independent advice and criticism to the project manager and project participants. This element was added very belatedly to GCES and never came to full maturation. The NRC committee fulfilled some functions of the advisory board but was not charged with giving constant operational advice to GCES and therefore did not provide all of the services that a true advisory board could.

As an adjunct to the use of an advisory board, all major study products, such as reports on project components, should be subjected to independent critique and review rather than remaining internal to the study group. Some review can be accomplished by publication in peer-reviewed literature, but other mechanisms for review also can be used for study components that are not appropriate for publication in full.

Achievements and Dissemination of Information

Another element of implementation is the organizational framework for dissemination and processing of data. Each large project such as GCES needs a general archiving and organizational system. Such a system was worked out by GCES through the use of the Geographic Information System and other computerized information storage systems. While not fully operational as of the end of GCES, the framework was correctly conceived.

Reference to Final Objectives

A successful ecosystem analysis requires constant referencing of individual project components to the project's final objectives. Retention of appropriate scope for a project is a constant responsibility of management and cannot be executed in one step at the beginning of the project. The manager of the project and the manager of the advisory group need to ask continuously each project component and each participant how specific kinds of data collection will come to bear on the evaluation of management options. When this question cannot be answered satisfactorily, resources should be redirected to other objectives that are more pressing.

Budgetary Continuity

Federally funded projects can be subject to particular budgetary uncertainty if they are supported as a marginal activity of a major agency. This was the case with GCES, which ultimately was continued without interruption for 13 years but in most years was without any secure basis for budgetary planning. Ecosystem analysis is inherently a multiyear activity, although the primary phases of most ecosystem studies will not require as much time as those of GCES. A sponsoring agency should require a study plan leading to specific useful outcomes in a specified period of time, with specified costs.

Once justified, these arrangements should be a priority within the agency, and the study manager should be responsible for producing complete study products and adhering to the budget. In contrast to this ideal, GCES often showed an ad hoc approach to time schedules, partly because GCES was left in fiscal limbo toward the end of each budget year and partly because there was no binding list of study products.

COMPLETION AND ANTICIPATION OF FUTURE NEEDS

Completion of an ecosystem analysis involves final synthesis and recommendations to management, archiving of study results and data for future use, and recommendations for selective additional studies or monitoring.

Synthesis

Ecosystem analysis is almost useless without some final synthesis and recommendations to management. Even so, this is the phase of analysis that is least likely to be completed satisfactorily. For example, GCES, as of its end in 1995, had not produced any synthesis above the single component level and thus in a sense failed to reach its final objective. While the BOR contemplates future preparation of a synthesis with post-GCES funds, the failure of GCES to produce a more synthetic outcome directly connected to management makes for a poor demonstration of the usefulness of ecosystem analysis to management. A study plan must incorporate a firm commitment to final objectives, including explanation or modeling of connections between ecosystem components under the influence of management.

Archiving for the Future

Given that environmental regulations are a constant and growing element of management for any given site, environmental studies should be regarded as antecedents of future studies rather than as isolated projects. All of the basic data should be archived in standardized formats, and special studies should be written up in ways that make them and their underlying data useful in the future.

Continuity into the Future

Managers of environmental studies frequently feel strong motivation to recommend extension of their work as they are nearing completion. Such requests are often viewed with skepticism by sponsors, who seek a definite end to the project. There is validity in both viewpoints. A large environmental study can be viewed much like a large construction project. A major investment is made initially to create the corpus of the environmental analysis, much in the same way the initial investment is made in the physical structure of a dam. To be useful, environmental analysis typically requires some sort of long-term continuity in the form of monitoring, which might be likened to the routine maintenance or operation of a dam following the major investment of construction. In the absence of some extension of effort following a major ecosystem analysis, the continuing validity of the analysis and the availability of expertise on the system will fade rapidly and undermine the original investment. In addition, new insights or operational changes may require revision, including new kinds of data collection on selected components of the system, if the analysis is to remain useful.

ACHIEVEMENTS OF GCES

Although the deficiencies of GCES were many, GCES can also claim numerous achievements, some of which relate to major expansion in our understanding of the Colorado River ecosystem between Glen Canyon Dam and Lake Mead, while others are of a more general nature (Table 11.1).

Important Discoveries of GCES

GCES made a number of basic discoveries that can be used as a basis for optimizing the operation of Glen Canyon Dam in ways that benefit biotic communities, recreation, and other resources. The studies of sediment transport, which in some ways were the most satisfactory component of GCES, showed that sand entering the Colorado River through the Paria and the Little Colorado rivers, and to a lesser degree other small tributaries, is sufficient to provide the mass of sand necessary for maintenance of beaches and backwaters along the Colorado River below Lee's Ferry. Prior to GCES, it was generally suspected that the amount of sand from these sources would

TABLE 11.1 Milestones of Achievement for the BOR Through GCES

1. Use of the ecosystem concept for redefinition of GCES scope.
2. Addition of senior scientist to GCES.
3. Manipulation of discharge as a means of studying ecosystem responses.
4. Recognition of the potential trade-off between power production and environmental benefits.
5. Recognition of the importance of long-term monitoring.
6. Inclusion of Native American tribes as cooperators.
7. Commitment to active management through controlled floods (beach-building flows).
8. Recognition of nonuse values.
9. Initiation of studies on multiple outlet withdrawal.
10. Creation of power resource studies involving external review.
11. Increase in extramural (nongovernmental) contracting.

be insufficient for this purpose. Therefore, GCES showed that management of sand is feasible below Lee's Ferry given the existing sediment supplies without augmentation from a slurry pipeline or other sources.

GCES also showed that controlled floods (now called beach-building flows) must be used to manage sand and debris (cobble and large rocks) in the canyon below Lee's Ferry. While the amount of sand entering the river is sufficient to maintain beaches and backwaters, it will not do so in the absence of occasional flood flows that are sufficient to lift sand from the bed of the river and over the tops of beaches and to scour backwaters so that they do not become filled with sediment. Beach-building flows are an ideal management tool because they present low environmental risk and cause the sacrifice of only small amounts of power revenues in that they need to last only a few days and need not occur every year. Beach building flows were set for spring of 1995 but were cancelled by the BOR due to legal concerns from the upper basin states. Another experimental flood flow is scheduled for spring 1996.

The GCES sediment studies also showed that moderation of ramping rate and particularly the downramp (decline of discharge) within the 24-hour cycle could offer substantial environmental benefits. The interim flows and subsequent preferred alternative of the environmental impact statement (EIS) incorporate moderated ramping rates that involve small losses in power revenue but create a flow regime that is more appealing for recreational use (rafting and fishing), less likely to cause stranding of trout, and less likely to accelerate loss of beach sand through the slumping that occurs during the downramp phase. Moderation in the extremes of discharge within a given day offers many of the same benefits and may enhance the biotic value of backwaters along the Colorado River.

The GCES showed that the humpback chub is, as previously suspected, almost entirely dependent on the Little Colorado River for its maintenance in the Colorado River between Glen Canyon Dam and Lake Mead. The intensive studies of humpback chub also showed that small populations are present near other tributaries, suggesting that a second population center might be established in the future. These studies were initiated late in GCES, however, and have not yet produced final conclusions. The effects of various operating schemes on the Kanab amber snail and the willow flycatcher of the riparian zone also are unclear at this time.

The definitive results of GCES primarily involve the management of sediment. While modest in number, these results are of great practical value and have led to the acceptance of new management schemes that will produce substantial environmental benefits with only modest loss of power revenues.

Recognition of the Need for Comprehensive Environmental Studies

Between 1983 and 1995, the BOR expanded the scope of GCES studies to realistic limits geographically and conceptually, accepted the ecosystem concept as the basis of the study and for interpretation of results, and acknowledged, during the EIS phase, the necessity of weighing power production and power revenues against environmental costs and benefits. These were all major advances in the management of the operation of Glen Canyon Dam. The dam is now managed according to an optimization concept that involves environmental, cultural, and recreational resources as well as power production and water management. While some of the management rationale is still not sufficiently backed by hard information, and the rationalization process itself is still deficient in some respects, the basic approach is sound and sets a framework that can be improved and refined in the future.

Adaptive Management and Long-Term Monitoring

The BOR and its cooperators have proposed adaptive management as a basis for managing Glen Canyon Dam in the future. This will be a marked contrast to the past management strategy, which was essentially static until interim flows were adopted. Adaptive management will require frequent review of information on all resources and adjustment of operations as needed to optimize benefits from the suite of resources that are affected by the operations of Glen Canyon Dam. Beach-building flows (controlled floods) will be a key feature of adaptive management. In addition, the use of a multiple outlet withdrawal structure might become a new element of adaptive management if comprehensive studies of this option prove to be encouraging. Thus, the adoption of adaptive management is an improvement in the management strategy for Glen Canyon Dam.

The EIS team, with encouragement from GCES and the NRC committee, has also specified that long-term monitoring of resources affected by the dam's operations will be important in the future. The BOR has authorized the development of a long-term monitoring plan, and the NRC committee sponsored a workshop on long-term monitoring in 1992 to assist in the development of this plan.

The commitment to long-term monitoring is essential as an adjunct to adaptive management. The environmental system below the Glen Canyon Dam is not static and thus will show numerous changes in the future that are responses to dam operations or to other events outside the realm of operations. These changes will be detected through long-term monitoring, and adaptive management will allow appropriate responses.

One difficulty with the BOR's commitment to long-term monitoring has been the absence of any specific monitoring plan that could be subjected to debate and criticism prior to its adoption. A draft plan was formulated by GCES following the NRC committee workshop in 1992. Although the NRC committee criticized the draft plan in some detail, no revised plan has yet appeared. For reasons outlined in Chapter 2, adaptive management must be served by a specific plan that evolves around needs for information rather than constituencies, political forces, and precedents.

External Expertise and Review

The BOR made a major administrative advance in appointing a senior scientist to assist the project manager of GCES. The scientist was drawn from outside the federal government and brought an independent perspective to GCES. In addition, the BOR approved selective use of the advisory board principle mentioned in Chapter 2, although this principle was never fully developed by GCES. These elements, which strengthen an agency-sponsored project by drawing on external expertise and promoting constructive criticism, are commendable and need to be extended in the future by greater use of external contracting and review.

A HOPEFUL VIEW OF THE FUTURE

While agencies of the U.S. government are notoriously conservative, the BOR has shown through GCES its ability to adapt to changing circumstances and new societal priorities. Despite the appearance in 1983 that the operations of Glen Canyon Dam would never be altered, the BOR has redirected the management of the dam in ways that take into account the many valuable amenities and resources of the Colorado River corridor below the dam. The GCES and the changes that have come about through preparation of the operations EIS have modernized and reformed resource management in the Grand Canyon region. While many problems remain to be solved, the basic elements of a responsive and enlightened environmental management system are in place at Glen Canyon Dam. The BOR has made a significant step in broadening its mission from purveyor of water to environmental manager. The lessons of GCES are, to a large extent, transferrable to other locations and could be the basis for a new era in the management of western waters.

REFERENCES

National Research Council. 1987. River and Dam Management. Washington, D.C.: National Academy Press.

Wilkinson, C.F. 1993. Crossing the Next Meridian: Land, Water and the Future of the West. Washington, D.C.: Island Press.

Appendix A

Biographical Sketches of Committee Members

William M. Lewis, Jr., *(Chair)* is professor and chair of the Department of Environmental, Population, and Organismic Biology at the University of Colorado, Boulder, and serves as director of the Center for Limnology at CU-Boulder. Professor Lewis received his Ph.D. degree with emphasis on limnology, the study of inland waters, in 1974 from Indiana University. His research interests, as reflected by over 120 journal articles and books, include productivity and other metabolic aspects of aquatic ecosystems, aquatic food webs, composition of biotic communities, nutrient cycling, and the quality of inland waters. The geographic extent of Professor Lewis's work encompasses not only the montane and plains areas of Colorado but also Latin America and Southeast Asia, where he has conducted extensive studies of tropical aquatic systems. Professor Lewis has served on the National Academy of Sciences/National Research Council's Committee on Irrigation-Induced Water Quality Problems and was chair of the NRC's Wetlands Characterization Committee. He is a member of the NRC's Water Science and Technology Board.

Garrick A. Bailey earned his B.A. in history from the University of Oklahoma and his M.A. and Ph.D. in anthropology from the University of Oregon. He is a professor in the Department of Anthropology and is director of the Indian Studies Program at the University of Tulsa. Dr. Bailey specializes in North American Indians, legal systems, cultural ecology, ethnohistoric methods, and social organization. He is a member of the American Anthropological Association, the Plains Anthropological Society, the American Ethnological Society, and the American Society of Ethnohistory.

Bonnie Colby is associate professor of agricultural and resource economics at the University of Arizona. Her undergraduate degree is from the University of California and her Ph.D. from the University of Wisconsin. Her research, teaching, and consulting focus is on the economics of water resources management and policy. She has authored over 40 publications in this area, including a number of journal articles and a book, *Water Marketing in Theory and Practice: Market Transfers, Water Values and Public Policy* (1987). In addition to her work on water reallocation, she has specialized in research on water quality, valuation of water rights and environmental amenities, and natural resource management in developing tribal and rural economies. Dr. Colby served on the NRC's Committee on Western Water Management.

David Dawdy received his M.S. in statistics from Stanford University. His professional experience has been with the U.S. Geological Survey from 1951 to 1976 as a research hydraulic engineer; as adjunct professor of civil engineering at Colorado State University, Ft. Collins, from 1969 to 1972; and as assistant district chief for programming of the California District, Water Resources Division, from 1972 to 1975. He has served on numerous advisory groups including NRC committees. From 1976 to 1980 he was chief hydrologist with Dames and Moore in Washington, D.C., and is currently a consultant in surface water hydrology.

Robert C. Euler is a consulting anthropologist specializing in the applied anthropology, archeology, ethnology, and ethnohistory of the American Southwest and Great Basin. As such, he conducts research in cross-cultural resources management, social and economic impact assessments, Indian legal claims cases, and archeological investigations, especially those related to environmental impacts. Dr. Euler is also adjunct professor of anthropology at Arizona State University, Tempe. In addition, he serves as tribal anthropologist for the Yavapai-Prescott Indian Tribe. Dr. Euler earned his B.A. and M.A. in economics from Northern Arizona University and his Ph.D. in anthropology from the University of New Mexico.

Ian Goodman earned his B.S. in civil engineering from Massachusetts Institute of Technology in 1977. Initially in his career he performed research at MIT, where he developed inputs to a policy-specific model of energy use for intercity goods movement. He began consulting in 1978 and was employed with several firms in the Boston area, working on various aspects of utility regulation and economics. He is now the principal of his own consulting firm, The Goodman Group, where his work includes assessing electric and gas resource planning, demand forecasts, supply options, and environmental effects. Mr. Goodman also evaluates conservation potential and cost effectiveness, program design, and utility demand-side management initiatives.

William Graf obtained his Ph.D. from the University of Wisconsin, Madison, with a major in physical geography and a minor in water resources management. He specializes in fluvial geomorphology, hydrology, conservation policy and public land management, and aerial photographic interpretation. He has served as consulting geomorphologist for the U.S. Army Corps of Engineers in a research and advisory role concerning the environmental impact assessment of flood control works at the Salt and Gila rivers in Arizona; and for Camp, Dresser, and McKee, Inc., for geomorphology and geology, and the state of Arizona for fluvial geomorphology. His research activities have emphasized fluvial geomorphology and the effects of human activities on streams; public land management, especially wilderness preservation, and rapids in canyon rivers; dynamics and recreation management; and the problems of heavy metal and radionuclide transport in river systems. Dr. Graf has published some 50 articles and book chapters on the impact of suburbanization on fluvial geomorphology; resources, the environment, and the American experience; and the effect of dam closure on downstream rapids. His books include *The Geomorphic Systems of North America*, *The Colorado River: Basin Stability and Management*, *Fluvial Processes and Dryland Rivers*, *Wilderness Preservation and the Sagebrush Rebellions*, and *Plutonium and the Rio Grande*. Dr. Graf is a member of the NRC's Water Science and Technology Board.

Clark Hubbs received his Ph.D. in biology from Stanford University in 1951. He joined the faculty of the University of Texas at Austin in 1949, became professor of zoology in 1963 and was the Clark Hubbs Regents Professor in 1989, and has been regents professor emeritus since 1991. He has served as chairman of biology (1974-1976) and chairman of zoology (1978-85). He was concurrently visiting professor of zoology at the University of Oklahoma (1973-1986) and on the faculty of Texas A&M University (1975-81). He has served as curator of ichthyology at the Texas Memorial Museum since 1975. He has received the Award of Excellence from the American Fisheries Society and the Lifetime Achievement Award from the American

Society of Ichthyologists. He has published more than 250 papers on aquatic biology. His research interests include distribution and speciation of fishes, hybridization of freshwater fishes, and environmental modification of freshwater fishes. Dr. Hubbs has a history of work with endangered fishes and now has a substantial program on predation of adults on their young.

Trevor C. Hughes acquired his Ph.D. in civil engineering from Utah State University. His professional experience includes teaching since 1972 at Utah State University in the Civil and Environmental Engineering Department; research experience as NDEA fellow at Utah State; associate professor of civil and environmental engineering, Utah Water Research Lab; and research scientist at the International Institute of Applied Systems Analysis, Austria. Since 1971 he has conducted research projects on the management of salinity in the Colorado Basin, drought management analysis and policy design, regional planning of rural water supply systems, economic analysis of alternative water conservation concepts, river system operational models, and modeling of urban water system demands.

Roderick Nash received an M.A. and Ph.D. in 1961 and 1964 from the University of Wisconsin. He specialized in American intellectual history under Professor Merle Curti. Before his appointment at the University of Santa Barbara in 1966, he taught for two years at Dartmouth College. Dr. Nash published the first collections of documents relating to environmental history, *The American Environment*, in 1968. His most significant recent work, *The Rights of Nature: A History of Environmental Ethics*, was published in 1989. A national leader in the field of conservation, environmental management, and environmental education, Dr. Nash has a special interest in problems relating to the wilderness and its preservation.

A. Dan Tarlock holds an A.B. and LL.B. from Stanford University and is currently Distinguished Professor of Law and Associate Dean for Faculty at the Chicago-Kent College of Law. He has practiced law in San Francisco and Omaha-Denver, and taught at the universities of Chicago, Indiana, Kansas, Michigan, Texas and Utah. He has written and consulted widely in the fields of water law, environmental protection and natural resources management. From 1987-1994, he was a member of the Water Science and Technology Board, and between 1989-1992 he chaired the Committee on Western Water Management Change, the report of which was published as *Water Transfers in the West* in 1992.

NATIONAL RESEARCH COUNCIL STAFF

Sheila David, a senior program officer at the Water Science and Technology Board, served as study director for this committee since its inception in 1986. On the staff of the National Research Council (NRC) since 1976, she has served as study director for a wide range of NRC projects including studies on coastal erosion, wetlands characteristics and boundaries, ground water protection, water reuse, and international studies concerning water supply management in Indonesia and the Middle East region.

Mary Beth Morris is a senior project assistant at the Water Science and Technology Board. She has been on the NRC staff since 1993 and has worked on WSTB studies including flood risk management, the use of treated wastewater on crops for human consumption, and valuing ground water. She holds a B.A. in politics from Randolph-Macon Woman's College.